Cuba's Socialist Economy Toward the 1990s

Cuba's Socialist Economy Toward the 1990s

Edited by
Andrew Zimbalist

Lynne Rienner Publishers • Boulder & London

Published in the United States of America in 1987 by
Lynne Rienner Publishers, Inc.
948 North Street, Boulder, Colorado 80302

and in the United Kingdom by
Lynne Rienner Publishers, Inc.
3 Henrietta Street, Covent Garden, London WC2E 8LU

© 1987 Pergamon Journals Ltd. First published as Volume 15, Number 1 (January 1987), of *World Development*. This edition published by arrangement with Pergamon Journals Ltd.

Introduction © 1987 Lynne Rienner Publishers, Inc.

Library of Congress Cataloging-in-Publication Data

Cuba's socialist economy toward the 1990's.

  Bibliography: p.
  Includes index.
  1. Cuba—Economic conditions—1959–
2. Agriculture and state—Cuba. 3. Industry and state—Cuba. 4. Public health—Cuba. 5. Central planning—Cuba. I. Zimbalist, Andrew S.
HC152.5.C824 1987   338.97291   87-12867
ISBN 1-55587-080-5 (lib. bdg.)
ISBN 1-55587-081-3 (pbk.)

Printed and bound in the United States of America

The paper used in this publication meets the requirements of the American National Standard for Permanence of Paper for Printed Library Materials Z39.48-1984. ∞

This book is dedicated to the memory of my sister, Michelle Zimbalist Rosaldo (1944-1981), who, among many other things, began to teach me the meaning of critical scholarship.

Contents

	Acknowledgments	vii
	Introduction: Cuba's Socialist Economy Toward the 1990s *Andrew Zimbalist*	1
1	Patterns of Cuban Development: The First Twenty-five Years *Andrew Zimbalist & Susan Eckstein*	7
2	Agricultural Policy and Development in Cuba *José Luiz Rodríguez*	25
3	Gender Issues in Contemporary Cuban Tobacco Farming *Jean Stubbs*	43
4	The Performance of the Cuban Sugar Industry, 1981-85 *Carl Henry Feuer*	69
5	Cuban Industrial Growth, 1965-84 *Andrew Zimbalist*	85
6	Development and Prospects of Capital Goods Production in Revolutionary Cuba *Claes Brundenius*	97
7	The Cuban Health Care System: Responsiveness to Changing Needs and Demands *Sarah M. Santana*	115
8	Worker Incentives in Cuba *Alexis Codina Jiménez*	129
9	Power at the Workplace: The Resolution of Worker-Management Conflict in Cuba *Linda Fuller*	141
10	Cuban Planning in the Mid-1980s: Centralization, Decentralization, and Participation *Gordon White*	155
11	Trade, Debt, and the Cuban Economy *Richard Turits*	165
	The Contributors	183
	Index	185

Acknowledgments

Researching Cuba is no easy matter. The labyrinth of hurdles one has to negotiate is well known to all who have tried to study post-1958 Cuba, given the prevailing disinformation, ideological cascade, and emotional overlay. One cannot successfully run the gauntlet alone. I am indebted to too many friends and colleagues for their intellectual and emotional support to acknowledge here. I would, however, like to single out the following individuals: Stuart Brown, Claes Brundenius, Carmen Deere, Anne Drabek, Susan Eckstein, Miguel Figueras, Nestor García, Lisa Genasci, Gretchen Iorio, Roger Kaufman, Sinan Koont, Arthur MacEwan, Juan Carlos Martínez, Jésus Molina, Lydia Nettler, Ernesto Ortega, Nola Reinhardt, Ariel Ricardo, José Luis Rodríguez, Charles Staelin, Armando Santiago, Jean Stubbs, Juan Valdés Paz, and Fidel Vascós. Finally, my profound gratitude goes to Jean and Harvey Picker for their financial generosity, intellectual stimulation, and personal encouragement.

Introduction: Cuba's Socialist Economy Toward the 1990s

ANDREW ZIMBALIST

Cuba has the oldest and most developed economy, except for China's, of the Third World socialist nations. As such, Cuba serves as an economic model, as well as a source of aid, guidance, and inspiration to many countries identifying with this expanding group. For this reason and because of Cuba's prominent international political trajectory, Cuban economic performance and structure have been the subject of considerable interest and continuing controversy.

In late 1974, one U.S. government report characterized the Cuban economy as being "on the verge . . . of constructing a socialist showcase in the Western Hemisphere."[1] By the 1980s, U.S. government portrayals of the Cuban economy had changed appreciably, either declaring the entire economy to be an utter failure or acknowledging significant achievements in the social sphere, but attributing the same to massive Soviet aid.

In part, these differing evaluations of Cuba's economy reflect the political prejudices of the various authors and the government they represent, and in part they reflect the generally poor state of existing scholarship on the Cuban economy. That is, without systematic treatment of the subject and in the absence of a scholarly consensus, politicians have been less constrained in constructing an interpretation of Cuba in accordance with their own political biases and interests.

Part of the problem here belongs to the scholars themselves, but an important part also belongs to (a) the paucity, until recently, of timely economic data from Cuba, (b) the lack of access to Cuban policy makers and economic institutions for investigators, and (c) the absence of scholarship by Cubans on the subject. With economic institutionalization, financial internationalization, and healthy economic growth since the early 1970s, however, the Cubans progressively have devoted more resources to the generation and dissemination of consistent statistical series, as well as to the scholarly investigation of their economy.[2] They also have gradually opened up, albeit cautiously, to foreign researchers. The last five years, then, have witnessed a more serious, detailed discussion and evaluation of the Cuban economy.

The present volume exploits this propitious scholarly development. It also includes chapters by Cuban economists analyzing their own economy. Although the standards of independent, critical analysis are not the same, Cuban economists increasingly are producing interesting and useful accounts and interpretations of their economy. It is important to establish a dialogue between Western development economists and their Cuban counterparts, and the pieces written by Cubans are included in that spirit.

This collection begins with an overview of Cuban development since 1959. It then proceeds from the specific to the general: it considers first the agricultural sector, then the industrial sector, next basic needs and labor issues, and finally central planning and foreign trade, aid, and debt.

The overview, written by Andrew Zimbalist and Susan Eckstein, describes the evolution of development strategy and records the changing economic performance over the first twenty-five years of Cuban socialism. It endeavors to recast the conventional interpretation of shifting development strategy by showing how both ideological and material considerations had an impact on policy formation. This reinterpretation is not only of historical interest, but can shed light on the current 1986/87 renewal of the debate concerning the appropriate use of moral versus material incentives in the construction of socialism. The reconsideration of incentives surely has an ideological component. It is less often recognized, however, that material incentives are severely limited in their usefulness in a non-market, shortage-ridden economy. Given their limitations, it is imperative for the effective functioning of centrally planned economies to develop, as well, moral and internal incentives (i.e., workers' internalization of goals through greater participation in goal formulation).[3] The limitations of material incentives become more pressing during periods of severe foreign exchange difficulties, such as Cuba is currently experiencing.[4]

The next three chapters deal with agriculture. José Luiz Rodríguez, one of Cuba's leading economists, analyzes the deficiencies in Cuban agricultural development prior to 1959 and sketches the post-1959 evolution of agricultural policy and structure, demonstrating leadership flexibility in accepting a sizable private sector and ingenuity in promoting new forms of organization.

Jean Stubbs analyzes the complex matrix of distinct organizational forms in agriculture, cultural patterns, and their impact on the growing economic role of women. She focuses her attention on the development of tobacco cooperatives in two different communities and concludes that cooperatives, potentially, can play a progressive social and economic role. Her treatment offers an insightful, firsthand account of the dynamic functioning of the new cooperative form that has come to dominate the private sector in Cuban agriculture.

Carl Henry Feuer discusses the central, yet changing, role played by sugar in Cuba's economy. He critically analyzes dependence on sugar and the role of Soviet price subsidies, as well as the dynamic linkages sugar has provided to the industrial sector since the revolution. Recent developments in labor policy and organization, the role of women, the growth of sugar byproducts and new technology, the introduction of agroindustrial complexes, as well as implications of present sugar policy for the future of Cuban economic development, are skillfully treated.

One important change in agricultural policy is not treated in these articles: the 15 May 1986 decision to abolish the free farmers' markets that were begun in 1980 and had become an increasingly important outlet for the sale of foodstuffs to the population.[5] From the leadership's point of view, however, these markets were being exploited by middlemen and price gougers who were profiting from the labor of others and generating excessive inequality in rural areas. Further, their activities engendered resentment and undermined the development of socialist consciousness. Finally, the farmers' markets were sustained by resource diversion away from the state sector. Some truck drivers for state farms, for instance, shortened their scheduled work day in order to transport produce from private farms to these markets.

The elimination of these markets, however, has presented a challenge to the state distribution system. The great majority of peasant producers should not be directly threatened by this change because they should be able to sell to the state the same excess production they had been selling to private wholesalers or taking themselves to the market. State procurement (*acopio*) prices for above-plan output are similar to those in the free markets for most commodities, and where they are lower they can be raised. The post-1982 20 percent sales tax on farmers' market sales is also avoided. The more pressing problem will be for the state procurement system to expand their transportation, distribution, and storage capacity sufficiently, in the short run, to be able to service all the producers previously handled through the peasant market system. Some difficulties and inconveniences will persist, but the leadership reported in early July 1986 that the process was under control, that most peasant producers had responded positively to the new measures, and that the prospects for agriculture in the near term were positive.[6] In its quarterly report of October 1986, the vehemently anti-Castro staff of Radio José Martí indicated that there were no reports of serious dislocations resulting from this new policy.[7] (The state-run parallel market network has expanded rapidly since the initial 1982 crackdown on farmers' markets. In 1985 sales on these markets grew 14 percent to 679.1 million pesos, or some 15 percent of total household consumption expenditures. Sales on these markets continued to grow during the first nine months of 1986, expanding by 42.8 million pesos overall or by 31.4 million, considering only foodstuffs.)[8]

The next two chapters deal with Cuban industrial development. The Zimbalist piece critically evaluates recent independent estimates of Cuban industrial growth and then, based on new data, offers its own estimate. It discusses price formation in Cuba and the effect of Cuban, non-market prices on growth measurements. The Zimbalist estimates are very close to (although slightly below) the official growth rates for the 1965–84 period and tend to lend heightened credibility to the official statistics suggesting strong growth since 1970. (Applying the Zimbalist methodology to 1985, the industrial growth for that year was 4.1 percent, and the growth rate probably slowed further in 1986 when the overall economy grew at lower than 2 percent in real terms. The slower growth in recent years is attributable largely to the direct and indirect effects of Cuba's foreign trade problems.)

The Claes Brundenius chapter is an in-depth look

at the capital goods branches of Cuban industry. Brundenius explains the evolution of Cuban policy toward these branches, as well as the basis for its achievements. His independent estimates suggest annual growth in excess of 15 percent for capital goods between 1965 and 1985. As he argues, capital goods provide the basis for mechanization, autonomous technological growth, and higher productivity throughout the economy. The success in these branches helps to explain the broader economic success in Cuba since 1970. Together, the chapters make clear that the industrial sector has become a dynamic and central part of the Cuban economy.[9]

The next three chapters discuss labor and social issues. The piece by Sarah M. Santana focuses on the Cuban strategy for health delivery and analyzes the bases of its well-known and remarkable success.[10] The present policy to develop neighborhood family physicians is seen, *inter alia,* as a function of existing health care structures, the changing role of women in society, and the emergence of a new generation of physicians. The decentralization of health care delivery implied by this policy has been recently the subject of much criticism in the Cuban press. Specifically, the problems of labor indiscipline have increasingly taken the form of workers presenting unwarranted medical excuses from doctors to absent themselves from work. The tensions engendered by decentralization in a centrally planned economy are thus manifested even in the most successful spheres of Cuban society.

Alexis Codina Jiménez outlines the evolution of the policy and practice regarding worker incentives. Using official documents, Codina is able to clarify the complexity of implementing material bonus schemes for workers in a planned economy and thus helps to anticipate the present reevaluation of the appropriate balance between moral and material incentives. Codina also discusses at some length the experiences with Cuba's newest workplace experiment, production brigades. The brigades, by decentralizing the locus of decision making, as well as the unit of account, offer the prospect for both greater worker participation and productivity. There is, of course, also a potential danger in the brigade experiment. With more independent financial units now functioning within each enterprise it multiplies severalfold the quantity of monthly reports, general paperwork, and meetings. The Cubans already complain about the excess of *burocratismo* and *reunionismo* that has accompanied efforts to rationalize their planning system.[11] Nevertheless, the brigades do offer the possibility of transforming social relations in Cuban enterprises and, given the stubborn problems with individual output norms and bonuses that have been thoroughly evaluated in recent public discussions, it would appear to place an even greater burden on alternative forms of motivating work effort and creativity.[12]

Linda Fuller, using personal interviews, public documents, and newspaper accounts, provides a careful and detailed look at one aspect of workplace social relations—grievance resolution. This proves to be a useful vantage point for evaluating the relative power of workers in enterprise management and for assessing the claim of greater worker rights under socialism. Fuller emphasizes the importance and subjective character of the local union in establishing the balance of power between management and the workers. She also treats the close relationship between shifting grievance procedures and regulations and the prevailing macroeconomic conditions.

The final two articles deal with Cuba's macroeconomy. Gordon White describes briefly the operation and reform of Cuban central planning. He analyzes the weaknesses of Cuba's young (post-1976) planning system and the systemic constraints it imposes on increasing substantial worker participation and improving efficiency. At the same time, White acknowledges the arguments in favor of strong central planning to promote economic development. Nonetheless, he looks for increasing use of markets and parametric controls as Cuba struggles to perfect its economic mechanism. Although this indeed may be the long-term direction for Cuban planning, since May 1986 the pendulum seems to have swung, temporarily at least, in the other direction. Insofar as further decentralization is pursued in the near term, it is likely to assume an administrative (e.g., production brigades or direct supply contracts between enterprises) rather than a market form (e.g., farmers' markets or substantive price reform). It is important to emphasize, however, that Cuba's current retrenchment follows roughly ten years of efforts toward decentralization. Given the experience with choppy reform paths in other centrally planned economies, such as the Hungarian retrenchment in 1971-72, it would be premature to say that Cuba has committed itself to a more centralized style of economic planning and management. The intermittent moratoria on reform processes appear to serve a necessary function in allowing each country to redress excesses, assess unexpected developments, and carefully chart its future course.

Richard Turits presents and analyzes the experience of Cuba's foreign sector. Regional and political patterns in Cuba's external trade, the persistence of Cuba's current account deficit, the management of its hard currency debt, and its special status within the CMEA are carefully explored. He argues that CIA estimates of Soviet aid employ a faulty methodology and are misleading: although it is falsely inflated by these estimates, Soviet aid is still very large and important to the Cuban economy. He identifies failures and successes, looming difficulties, and positive prospects for Cuba's foreign sector over the coming years. Turits's analysis anticipates many of the critical problems that have come to dominate Cuba's foreign trade and debt situation in 1986-87.

In 1986, Cuba's hard currency export earnings were severely reduced by a number of factors. First, reexports of Soviet crude oil, which had come to account for 40 to 42 percent of hard currency earnings over 1983-85, were acutely affected by the drop of world oil prices from approximately $30 to $12–$15 per barrel. In 1986, the value of these exports was projected to fall to only 26 percent of (greatly diminished) hard currency exports. Second, average sugar prices on the world market continued to be very low, beginning a slow recovery from their 1985 average of 4.04 cents per pound. Third, the rapid devaluation of the dollar vis-à-vis the currencies of the other OECD countries cost Cuba an estimated 150 million pesos in international purchasing power. Since the prices of most of Cuba's hard currency exports (e.g., sugar, petroleum) are denominated in U.S. dollars and Cuba does not trade with the United States, Cuba's hard currency import prices denominated in yen, marks, francs, etc., rose much more rapidly than its export prices. Debt service payments were also made more difficult by this phenomenon since Cuba's hard currency debt is largely denominated in these appreciating currencies. Fourth, following the devastating effects of Hurricane Kate in November 1985, Cuba has been hit by a serious drought. Rainfall in 1986 was 35 percent below the historical average. Among other things, this drought is held to be mostly responsible for an estimated shortfall in the 1986-87 sugar harvest of one million tons. Fifth, the Third World debt crisis has brought growing protectionism, and, consequently, Cuba's trade balance with the underdeveloped world has changed from a strong surplus at the end of the 1970s and the beginning of the 1980s into a growing deficit. Sixth, the Reagan administration has significantly tightened the blockade against Cuba. Much of this impact was felt prior to 1986 (e.g., reduced nickel exports and elimination of U.S. tourism), but additional measures regarding joint ventures in Panama, dollar transfers, and further travel restrictions were imposed in 1986. Seventh, Cuba has encountered increasing difficulty in rescheduling debt payments and obtaining new credits.

These factors led to an almost 50 percent current dollar reduction in Cuba's 1986 hard currency earnings. Although relative price movements during the first quarter of 1987 have been to Cuba's advantage, the severity of the situation is indicated by the following figures. In 1984—not a banner year—Cuba had $1.5 billion of hard currency available from exports and net credits. In 1986 dollars this sum represents approximately $2.0 billion in purchasing power. However, in 1986, hard currency availability was between $600 and $700 million, and a similar range was anticipated in the economic plan for 1987. This circumstance provoked Cuba to delay interest payments on medium- and long-term debts in the summer of 1986 and to make only sporadic service payments since. Further, at the end of 1986, the Cuban government introduced an extensive package of austerity measures designed to economize on foreign exchange. Real Gross Social Product in 1987 is planned to grow between 1.5 and 2.0 percent, similar to 1986 but considerably below the 1981-85 annual average of 7.3 percent.

There is little question that 1986-87 are years of austerity, reexamination, and change for the Cuban economy. It is also true, however, that austerity in Cuba does not have the same social or economic meaning as elsewhere in Latin America. There is no reason to expect unemployment, hunger, or falling output in Cuba during this difficult transition period.

Altogether, the essays in this volume offer a balanced, up-to-date discussion of Cuba's economic model and performance. It is important to underscore the fact that the Cuban model is in process. Continuing debate and experimentation have characterized the Cuban economic experience for twenty-seven years, and there is no indication that the Cubans have grown complacent either with their achievements or their institutions. If the articles contained herein are successful, they will have conveyed to the reader a sense of the effervescence and evolving quality of the Cuban political economy. They will also make clear that Cuban economic reality is considerably more complex than has been suggested by the simplistic and ideological characterization of many politicians and some academics.

NOTES

1. U.S. Senate Staff Report, Committee on Foreign Relations (1974), p. 1.

2. On the rapid development of the economics profession in Cuba and its professional journals, see Fitzgerald (1985), Chap. 8.

3. This argument is elaborated in Zimbalist (1984), Chaps.

1 and 7 and Brus (1973), Chap. 3. Also see Kornai (1980). Cuban difficulties in applying material incentives have been candidly reported in the press during the present discussions and reevaluation of economic policy. See, for instance, the lengthy articles concerning the *reuniones de empresas* (enterprise meetings) at the provincial level in *Granma* (15 July 1986), pp. 3–6.

4. In 1985, 42 percent of Cuba's hard currency earnings came from the reexportation of Soviet petroleum. The precipitous drop in world oil prices is costing Cuba over $200 million per year in lost earnings. Together with the low level of sugar prices (below six cents a pound in late July 1986) and the reduced 1986 sugar harvest occasioned by the ravages of Hurricane Kate, Cuba is experiencing severe cash flow problems and did not meet her scheduled debt principal or interest payments on 7 July. It has been reported that, in response to the falling commodity prices, the 1986-90 trade and aid agreement between the Soviet Union and Cuba increases the "direct aid package to Cuba by $3 billion, 50 percent more than in the previous five-year period" *Wall Street Journal* (24 July 1986), p. 27. The 50 percent increase itself was reported earlier in the Party newspaper, *Granma* (11 April 1986), p. 1.

5. Despite its growing role in 1985, according to one study, the share of free peasant market sales in total retail sales in Cuba was only 1 percent. Cf. Cepal (1986), p. 33.

6. See, for instance, the extensive discussion of this issue in the report on the Tenth Assembly of Popular Power held in early July 1986, *Granma* (5 July 1986), pp. 2–6. Also see Castro's closing speech at the Second National Meeting of Agricultural Production Cooperatives, reprinted in the *Granma Weekly Review* (1 June 1986), pp. 1–3.

7. Staff of Radio José Martí, *Cuba: Quarterly Situation Report* (Washington, D.C., October 1986), III., ii, pp. 3–8.

8. Banco nacional de Cuba, *Informe Económico*. (March 1986), p. 11; and, Banco nacional, *Quarterly Economic Report*. (September 1986), p. 7.

9. The importance and share of the industrial sector in Cuba's economy has been the subject of a recent scholarly debate. Mesa-Lago, by misinterpreting the impact of turnover taxes, argues that the industrial sector's share in Gross Social Product has fallen since the early 1960s. In fact, the share of industry has grown appreciably. Cf. Brundenius and Zimbalist (1985a and 1985b) and the Comité estatal de estadísticas (1985), pp. 83c, 95.

10. There have been some recent efforts to challenge Cuba's success in the health area. Eberstadt (1984) makes such an effort by, *inter alia*, misrepresenting a study of the National Academy of Sciences, overlooking the 1965 change in the definition of infant mortality from the Spanish (ignoring deaths in the first twenty-four hours of life) to the Western (counting all deaths from the first breath), and ignoring the severe underregistration of infant deaths prior to 1959. The quarterly report on Cuba by the Office of Research and Policy of the U.S. government funded Radio José Martí points to the increase in infant mortality from 15.0 per 1000 in 1984 to 16.8 in 1985, and suggests that, because of systemic factors, this increase will become a long-term trend. It presents no evidence regarding the alleged systemic factors; the infant mortality rate for 1986 actually fell to a record low of 13.6.

11. Notwithstanding its pretensions to decentralizing and increasing efficiency, the SDPE has engendered new bureaucratic entanglements. At the meeting to analyze enterprise management in Havana and City of Havana provinces held on 25–26 June 1986, Marcos Portal, the minister of basic industry, reported that in 1973 there were 90,000 administrative personnel in the country while, in 1984, there were 250,000—an increase of 178 percent. Portal's statement was cited in *Granma Weekly Review* (6 July 1986), p. 2.

12. See the sources in note 4. On the early realization of some of this potential, also see Veiga (1986).

REFERENCES

Banco nacional de Cuba, *Informe económico, (March 1986)*.
———, *Quarterly Economic Report*, (September 1986).
Brundenius, Claes, and Andrew Zimbalist, "Recent Studies on Cuban Economic Growth: A Review," *Comparative Economic Studies*, Vol. 27, No. 1 (Spring, 1985a), pp. 21–46.
———, "Cuban Economic Growth One More Time: A Response to 'Imbroglios'," *Comparative Economic Studies*, Vol. 27, No. 3 (Fall, 1985b), pp. 115–132.
Brus, Włodomierz, *The Economics and Politics of Socialism* (London: Routledge and Kegan Paul, 1973).
Cepal, *Notas para el estudio económico de América Latina y el Caribe, 1985: Cuba* (Mexico, julio 1986).
Comité estatal de estadísticas, *Anuario estadístico de Cuba, 1984* (Habana, 1985).
Eberstadt, Nick, "Literacy and Health: The Cuban 'Model'," *The Wall Street Journal* (10 December 1984).
Fitzgerald, Frank, *Politics and Society in Revolutionary Cuba: From the Demise of the Old Middle Class to the Rise of the New Professionals* (unpublished ms., 1985).
Kornai, Janos, "The Dilemmas of a Socialist Economy: The Hungarian Experience," *Cambridge Journal of Economics*, No. 4 (1980).
Staff of the Radio José Martí Program, *Cuba: Quarterly Situation Report* (Washington D.C., October 1986).
Theriot, Lawrence, U.S. Department of Commerce, *Cuba Faces the Economic Realities of the 1980s* (Washington D.C.: USGPO, 1982).
U.S. Senate, Committee on Foreign Relations, Staff Report, *Cuba* (Washington D.C.: USGPO, 1974).
Veiga, Roberto, "Discurso de clausura del encuentro nacional de la CTC para examinar la labor de las brigadas permanentes e integrales," *Granma* (30 enero 1986).
Zimbalist, Andrew (ed.), *Comparative Economic Systems: An Assessment of Knowledge, Theory and Practice* (Boston: Kluwer-Nijhoff, 1984).

·1·

Patterns of Cuban Development: The First Twenty-five Years

ANDREW ZIMBALIST & SUSAN ECKSTEIN

1. THEORETICAL INTRODUCTION

This paper attempts to reinterpret the evolution of Cuban development strategy and institutional change. It is hoped that this reinterpretation will shed new light on previous experience and policy decisions as well as on Cuba's economic prospects for the second half of the 1980s.

As a Third World country committed to socialist values and socialist organizing principles, the success of the Cuban revolution must be judged in terms of the extent to which (a) the country's economic base has expanded and its product redistributed to benefit previously deprived rural and urban groups and (b) the populace partakes in political and economic decision making.

Economic growth, other things equal, is positively correlated with the level of investment. Investable funds or surplus, in turn, originates in savings or loans. A central issue for the Castro government, therefore, has been its policy with respect to surplus generation and allocation.

Sources of surplus for economic expansion can be external or internal. External sources include trade surplus, foreign grants, loans and investment. Cuba's capacity to generate capital through external sources therefore must be understood within the context of CMEA (the Soviet trade bloc) dynamics and Cuba's particular status within CMEA, as the Soviet Union has come to be Cuba's principal source of aid and trade. However, since the Soviet bloc is not self-contained, Cuba's capacity to generate surplus through external sources also is shaped by Western trade and financing. With sugar remaining the island's principal source of export earnings, in its trade both with the Eastern and the Western blocs, its ability to generate surplus through trade continues to be shaped by volatile world sugar prices.

External sources of surplus may entail costs which limit their desirability to a government. They are desirable to the extent, *inter alia*, that interest charges on loans are reasonable and terms are not imposed which seriously restrict how the loans are used. Yet even when external financing meets these conditions, their domestic economic effect may not be unambiguously positive. Loan repayments often compel a government to orient production toward export. When a country cannot control the price of its exports, it may be forced to export at little or no profit. The extent of dependence on external sources of capital accumulation is reflected, for one, in the size of the foreign debt and its service payments, particularly in relation to the value of exports or the national product. The debt (and debt service) export ratio is particularly revealing as export earnings are the means by which most outstanding loans are repaid. Although a foreign debt in the short run may merely be indicative of a country's capacity to draw on external capital sources to stimulate domestic accumulation, one that increases over time will compel a diversion of resources in the future from domestic use to loan repayments.

Domestic bases of surplus accumulation hinge on savings. Domestic savings can be augmented either by sacrificing current consumption or by expanding output through the fuller use of existing resources (and saving part or all of the additional output). The latter method is, of course, politically more palatable. A fuller utilization of resources can be accomplished by putting unemployed labor or capital resources to work, by reducing waste, bottlenecks or resource underemployment, and by using existing resources more intensively.

Decentralization and democratization of decision making will shape how effectively resources are used as well as their allocation. Meaningful decentralization of decision making in a centrally-planned economy, however, can occur only if resources are available to decentralized units. For instance, for enterprises to exercise discretionary power over the use of their retained profits, the producer or consumer goods they desire must be available. If a reasonable variety of resources is to be available for decentralized purchase, there must be a generalized pattern of above plan (quota) production by economic units; such a phenomenon is known as "slack" planning, as opposed to "taut" planning where resources of production units are strained in order to meet full capacity level targets. Alternatively, slack can be planned and state warehouses can hold free reserves of inputs and outputs for extra-plan purposes. If slack is centrally controlled in this manner, however, much of the impulse toward decentralization will likely be frustrated.

Slackness can be built into a planning system over time by the absolute reduction of planned output targets, by having targets increase at a slower rate than actual production or production capacity, or by the expansion of non-state production. Since generalized reduction of planned targets is an improbable strategy, the slack necessary for meaningful decentralization rests with fuller use of existing capacity or the expansion of capacity. The latter requires either current sacrifices in consumption, saving out of increased output resulting from greater capacity utilization, or foreign capital.

The more repressive the regime, the more capital accumulation can rest on politically unpopular consumption sacrifices. The less repressive the regime, the more likely that slack will hinge on motivating workers to increase their productivity and (relatedly) permitting and encouraging managers and bureaucrats to introduce organizational changes that reduce waste. The more producers identify with the goals of their places of employment, the greater their motivation to work. Such identification is maximized when producers are involved in the process of goalsetting and implementation. Worker participation in decisionmaking accordingly contributes to the fuller use of production capacity. Resources are thereby generated both for capital accumulation and for the slack that makes for effective decentralization.

An evaluation of the costs and benefits of the different sources of surplus generation provides for a new understanding of why revolutionary policies have changed over the years, even when the policies might seem to be inefficient and economically unjustifiable from the point of view of output per unit of input. Such evaluations also help us understand why policies that appear, publicly, to have been guided primarily by ideology have been grounded in particular accumulation strategies. For this reason, the intent of this article is primarily interpretive: to provide a revisionist analysis of Cuban development since 1959.

It should be kept in mind that while the basis of surplus generation may be either domestic or foreign in origin, these sources interact in practice. We will show below how domestic and foreign sources of surplus accumulation have shaped the organization and performance of Cuba's political economy. The different mix of sources, and their effects, will be examined chronologically.

2. THE FIRST DECADE

The economy experienced an immediate production spurt under Castro due to improved use of installed capacity. However, the further expansion of the economy necessitated additions to productive capacity, which the government sought to realize via machinery and equipment imports. Since exports did not concomitantly expand, the country developed a large deficit in its balance of payments. As a result of the deficit, the government had to subordinate production for domestic consumption to production for export. It accordingly restructured production and modified the principal basis of rewarding labor. The export emphasis required such heavy centralization of decisionmaking that a political and economic crisis ensued. While the organizational and labor policy changes appeared to be ideologically grounded, it will be shown that there was a material base for the reforms as well. They addressed an underlying and preexisting accumulation crisis; they were not, as is generally assumed, the sole cause of the crisis.

Upon assuming power, the new government introduced redistributive and stimulative economic measures that put idle resources to work and contributed to the immediate expansion of the island's productive capacity. Prices were frozen, taxes were cut, rents were reduced, luxury imports were choked off, industrial profits were redistributed and a land reform was implemented that gave sharecroppers, tenant farmers, and squatters land rights and turned large estates into state farms or cooperatives (the early cooperatives were subsequently transformed into state farms). Meanwhile, employment expanded and full utilization was made of existing plant facilities. These policies, among others, bolstered demand, and, in turn, production. Between 1959 and 1961 agricultural output grew by an estimated 9% and between 1958 and 1961 manufacturing output expanded by approximately 8.3%[1].

By 1962, however, several problems set in. Easy economic growth with existing plant capacity was exhausted and further expansion required additional capital investment. Yet several factors limited the government's ability to accumulate the surplus that such investment required internally. First, the government had allocated large monies to social expenditures, both to train a new cadre of skilled personnel to replace the professionals, engineers and technicians who had emigrated, and to improve educational opportunities for, and the health and welfare of, previously deprived groups. While such social expenditures may contribute to capital accumulation in the long run, in the short run they come at its expense. Second, continued US threats and acts of sabotage pressed the government to divert resources from productive investment to defense. Third, the lack of a sufficient industrial infrastructure required extensive investment in electricity, roads, port facilities and the like before devoting new resources to expanding production capacity.

Meanwhile, surplus formation through foreign trade declined. On the one hand, the imposition of the US embargo caused capital input costs to rise exceptionally fast. The Soviet Union became Cuba's only viable source for most strategic imports. Yet Soviet equipment was more costly and inferior in quality to US equipment and Soviet imports entailed much higher transport costs. Also, machinery breakdowns had to be replaced with entirely new Soviet equipment, as the US embargo prevented Cuba from being able to purchase spare parts for prerevolutionary US equipment. On the other hand, export earnings fell as the country deliberately oriented production "inward." Consequently, by 1962 Cuba experienced a sizable balance of payments deficit.[2] The 1962 balance of payments crisis compelled the Castro government to subordinate production for domestic consumption to production for export, the exact opposite of its initial strategy. Because the Soviet Union, and to a lesser extent Eastern European countries and China, agreed to large long-term sugar contracts at stable, above world market prices, the government came to view exports not only as a source of revenue for reducing its foreign deficit but also as a source of surplus accumulation. Production oriented toward the domestic market did not offer a viable alternative source of capital accumulation in the short run, for reasons specified above, and it could not offer a solution to the balance of payments crisis; the latter could only be resolved through export promotion.

With the shift in emphasis, Cuba came to increase its dependence on sugar for trade, and exports came to assume a more important role in the economy. The big sugar push of the 1960s, however, failed to achieve the desired results. Sugar output rose from its 1963 low to a record high in 1970, but it did not rise annually as planned. Moreover, due to falling world prices, sugar export earnings did not rise proportionally with the volume of output and costs. Whereas long-term contracts with Soviet bloc countries shielded Cuba somewhat from the price volatility of the commodity, Soviet bloc purchases accounted for approximately 55% of the island's sugar exports by volume in the latter 1960s.[3] Most payment for sugar was in kind, calculated at inflated CMEA prices.[4]

The export strategy failed also because of its effect on domestic productivity. The overall growth rate of the economy dropped to about 0.4% a year between 1966 and 1970 (see Table 1). For one, the productivity of urban labor mobilized to help in the "big sugar push" was low and productivity in other sectors faltered as labor was diverted to sugar related activity. Therefore, the policies of the late 1960s adversely affected domestic as well as foreign sources of capital accumulation.

Second, the restructuring of the domestic economy and the new labor incentives introduced at the time of the sugar maximization program also had negative effects on productivity. During the 1960s the state nationalized progressively more of the economy, including most of agriculture in 1963 and commerce in

Table 1. *Official growth rates of Gross Social Product (GSP) average annual rates at constant prices*

	1962–65	1966–70	1971–75	1976–80	1981–85
GSP	3.7%	0.4%	7.5%	4.0%	7.3%
GSP per capita	1.3%	−1.3%	5.7%	3.1%	6.4%

Sources: Calculated from Mesa-Lago (1981, p. 34); Banco Nacional de Cuba (1984, p. 17); Comité Estatal de Estadísticas (1981, p. 67; 1982, p. 98; 1984, pp. 54, 95); CEE, *La Economía Cubana*, 1982, p. 17 and 1983, p. 18; *Granma Weekly Review* (16 February, 1986, p. 2); *Cuba: Indicadores Económico-Sociales y Políticos (1981–1985)* (Habana: Editora Politca, 1986, p. 1).
Note: According to the State Statistical Committee these official rates represent constant price or real rates of growth. Officially, recognized inflation (implicit GSP deflator) equalled 0.86% per year between 1976 and 1980, 11.7% in 1981, 1.2% in 1982, 1.2% again in 1983, and −0.04% in 1984. Although the standard GSP series for the period 1968–80 is given in the current prices due to the presence of price increases primarily in the commercial sector (in turn largely a result of the incorporation of turnover taxes in producer prices in this sector following the wholesale nationalization of small businesses in 1968), the numbers in Table 1 are, according to the State Statistical Committee, adjusted for price increases (possibly excepting the years 1968 and 1969) and hence represent real rates of growth. For an evaluation of the official Cuban growth statistics as well as independent estimates, see the debate between Mesa-Lago and Pérez-López, on the one hand, and Brundenius and Zimbalist, on the other, published in the journal *Comparative Economic Studies*, Spring and Fall issues, 1985; also see Zimbalist (1987).

1968. The 1963 agrarian reform nationalized all farms over 67 hectares (unless exceptionally productive), leaving only about 30% of the farmland and 30% of the agrarian labor force in the private sector. Accordingly, the state increased its capacity both to appropriate agricultural surplus and to determine how farm production was organized. Prior to the second reform, the government had difficulty expanding its revenue through taxes, and farmers did not necessarily produce goods that maximized state accumulation. They concentrated on production for the domestic market. Although numerous commentators have stressed that the 1963 agrarian reform was economically irrational, in that productivity on previously nationalized holdings was lower than on remaining private properties and that most state farms operated at a loss,[5] the reform was consistent with the government's strategy of promoting surplus accumulation through exports. Once in possession of the land, the government could divert land previously dedicated to products for the domestic market to sugar for export. Even if productivity dropped and even if per unit sugar costs increased (both of which proved to be the case), if trade revenue could thereby be generated the state's most pressing economic concern could be addressed. And indeed, between 1963 and 1970, sugar output increased while other farm crop production decreased, the opposite of farm output trends during the initial import-substitution period.[6] The extent of state-owned land devoted to sugar production increased by about 38% within one year of the reform, and it continued to rise for the remainder of the 1960s.[7]

Following the expansion of the state sector, the government modified the organization of production and labor relations. While the changes were publicly portrayed and widely believed to be hastening the transition to communism, in fact, they were consistent with the state's concern at the time with maximizing accumulation through both domestic savings and exports. For one, beginning in 1964 the state consolidated newly expropriated farms and it centralized control over local farm activities. Edward Boorstein, who worked for the Cuban government at the time, claimed the Cuba's heavy dependence on exports required centralized, autocratic control over production.[8] The reorganization, therefore, may well have been designed to maximize surplus from trade as well as to further socialize relations of production. Second, the government mobilized urban volunteers — or, in Mesa-Lago's terms, unpaid urban labor — for seasonal agricultural tasks, not solely to abolish distinctions between manual and "intellectual" labor and to emphasize moral over material work incentives, in accordance with communist principles, but also to maximize sugar earnings while minimizing direct production costs. According to available estimates, in 1968 about 15 to 20% of the people involved in agricultural production had been transferred from other sectors of the economy, and in 1970 more than one-third of the labor force worked part-time in agriculture.[9] Had the government relied on monetary incentives to attract urban labor to agriculture, the pay would have had to be high both because urban wages were then much higher than farm wages and because city jobs were considered preferable. Third, the government deprived state farm workers of small plots in the late 1960s previously alloted to them not only because the workers allegedly lavished too much attention on the plots but also because on the private parcels agriculturalists tended to produce goods for domestic consumption, not for export as on the state holdings.[10]

In the late 1960s the government also extended its control over production on remaining private farms to the point that private ownership lost much of its traditional meaning. It pressured private agriculturalists to sell all their produce to the state at low prices, it encouraged them to incorporate their properties into the state sector and to work part-time on state land, and it discouraged them from hiring labor. Furthermore, it attempted to induce private sector collaboration by making membership in the prestigious Advanced Peasants' Movement contingent on collaboration with state plans and production exclusively for the state. Since moral suasion was not sufficiently compelling to gain adequate cooperation from independent producers, the government began to rent private holdings. It accordingly expanded the sugar acreage under its control. Then, it lowered rents below the amount judged essential to cover necessities, in order to induce peasants not only to turn the land over to the state but also to work for the state.[11]

In constricting the activities of small farmers and eliminating the use of private plots by workers on state farms, the government sacrificed maximum short-run production and profit within *individual* units for maximum *economywide* production for export. Remaining private farmers had concentrated on production for domestic consumption rather than for export. They did so because production of the fruits and vegetables domestically consumed was labor-intensive and because they profited more from producing for the local market. While the government obligated private farmers to sell their produce to state collection agencies at low prices, a black market evolved in which large earnings could be made. Indeed, some of the wealthiest Cubans at the time were private farmers.[12] Therefore, the state here too had material and not merely ideological reasons for regulating private agricultural activity: its economic concern centered on maximizing export-based accumulation at the expense of total agricultural output. Sugar was its most internationally marketable farm commodity.

The organizational changes during this period were less dramatic in industry than in agriculture, as industry had been almost entirely nationalized during the first two years of Castro's rule. Until 1968 performance in the sector was more impressive than in agriculture. Between 1959 and 1968 output of slightly more than one-quarter of the major industrial items increased at least 100%. The increases were in basic industries and natural resource refining as well as in light industries. However, in 1969, industrial production contracted, even in the sectors that previously had performed well.[13] The drop in output in part resulted from the extreme emphasis on exports, as the economy mobilized for the 1970 10-million-ton sugar target. However, it undoubtedly also resulted from organizational changes initiated during the late 1960s. The government and Party centralized control over labor, reduced labor participation in enterprise and economywide decisionmaking, greatly restricted — as detailed below — material incentives, expanded the sphere of free goods, and ceased to elaborate either government budgets or nominally comprehensive economic plans.

The subordination of farm and industrial production for domestic consumption to production for export (along with emphasis on infrastructural development, health and education) was part of a general strategy to maximize capital accumulation. Accordingly, the government increased its overall investment ratio. Gross investment, as a share of national income (Gross Material Product) rose from 16.9% in 1962, to 25.3% in 1967.[14] To increase its investment capacity, the government also demonetized the economy and restricted wage earnings and material consumption. Exceptionally productive workers and workers who did overtime ceased to receive financial compensation beyond their basic wage. Wages of all but the poorest workers in the state sector were cut after having been increased earlier in the decade. Average wages for state workers declined 12% between 1966 and 1971, after having increased 3.5% between 1962 and 1966.[15] Only in the poorly paid agricultural sector did average earnings rise in the late 1960s.

The shift in the reward structure was justified ideologically, in terms of the communist principle of "from each according to his/her capacity, to each according to need." However, the government's concern with capital accumulation maximization calls for a revisionist interpretation of the official ideology and policy emphasis at the time. The government had material as well as ideological reasons for emphasizing equality, moral incentives, the combining of manual with non-manual labor, and the subordination of private to state sector activity.

With production for export, high investment ratios and low initial consumption standards, the availability of consumer goods and services during the late 1960s was insufficient for material incentives to be effective. To offer a worker higher wages in return for working harder and producing more could only be meaningful if the worker were able to convert his or her extra income into extra consumption. Given the extensive goods shortages at the time engendered by the accumulation strategy and adherence to egalitarian norms, there was no choice but to rely on non-material incentives to motivate work.

Mesa-Lago (1978), for example, errs in taking ideology at face value; he argues that government policies faltered in the late 1960s because they were premised on excessive idealism — so-called Sino-Guevarism — and non-pragmatic considerations.[16] While idealism contributed to the economic problems of the time, the strategy had a material base. The strategy in theory enabled the government to promote both trade and domestic savings and investment, and in so doing to address the balance of payments crisis. However, unfavorable world market prices — which fell to below two cents per pound for raw sugar between 1966 and 1968 — made the export strategy unviable. Moreover, some of the investments proved unwise while the payoff of others was not immediately forthcoming. The domestic labor force, in turn, was disinclined to accept the material sacrifices that resulted, particularly given the absence of significant worker participation and internal incentives. Consequently, the policies, and their ideological justification, contributed to a political and economic crisis. Although the state espoused a proletarian ideology, its concern with capital accumulation conflicted with the immediate interests of workers to improve their level of well-being and control over work. Workers resisted the rapid "push toward communism."

Labor expressed its dissatisfaction with the policies of the late 1960s more in the form of footdragging than political defiance. A time-loss study conducted in 1968 of more than 200 enterprises, for example, revealed that one-quarter to one-half of the workday was wasted due to overstaffing and poor discipline.[17] Absenteeism reached as high as 20% of the labor force following the 1970 sugar harvest.[18] The details of the productivity decline in the latter 1960s are not fully known but interpolated data suggest that the value of output per unit of labor input declined in agriculture, construction, and communication, while increasing in industry, transportation, and commerce.[19] Some of the registered increase in the latter two sectors may reflect price increases. Interestingly, according to this data, productivity improved in two sectors that experienced wage cutbacks and it decreased in the one sector that on average benefited from wage raises, suggesting that the material reward system in effect at the time did not in itself account for the economic crisis. Productivity declines were probably also rooted, inter alia, in the transfer of workers from private commerce to the state sector with the nationalization of small businesses in 1968; the seasonal mobilizations of non-agricultural labor for sugar-related tasks;[20] and the apparent deterioration in worker participation in decisionmaking (also partially a consequence of the accumulation strategy).

3. THE SECOND DECADE AND BEYOND

The political and economic crisis resulted in a shift in strategy in the 1970s. The government lowered its growth objectives while shifting its growth strategy, and it introduced various measures to rationalize the economy. Since it perceived the problems of the late 1960s to be based also in excessive political and economic centralization and bureaucratization, the leadership opened channels for participation "from below" in the workplace and on a territorial basis. The changes, together with technological improvements and propitious international conditions, had such a positive impact not only on consumption but also on surplus accumulation that sufficient resources and slack became available to implement economically meaningful and politically viable economic decentralization. However, later in the 1970s foreign sources of surplus deteriorated dramatically, to the point that the government once again faced serious trade imbalances and a burdensome foreign debt. The deep international recession of 1980–82, the return of very low sugar prices, and heavy debt service obligations have converted the external sector from a surplus producer to a surplus consumer. Nevertheless, the provisions for planned, subsidized trade with CMEA, along with ongoing success in import substitution, payoffs from earlier investments, and improvements in internal economic organization yielded remarkably strong economic growth again in the early 1980s. Changes in the organization of the domestic political economy and in the external sector, and their effects on surplus accumulation, will be discussed, in turn, below.

(a) *The restructuring of the domestic political economy*

As summarized in Table 1 above, the annual growth rate of Gross Social Product at constant prices averaged 7.5% during the first half of the 1970s. Industry, especially the capital goods sector, played a key role in the expansion.[21] A domestic resource base evolved around agriculture (which included production of harvesters, fertilizers, and pesticides) and fishing. Payoffs from previous investment and a better educated labor force undoubtedly contributed to the impressive economic performance of the early 1970s, but so too did domestic organizational changes that resulted in improved use of existing resources and propitious international conditions that enhanced external bases of surplus accumulation.

Organizational changes first centered in the workplace, then in the polity and economy at large. They served to broaden mass participation, particularly at the local level. The concern with organizational efficiency and domestically-based capital accumulation also resulted in the implementation of a new system of economic planning and management in the latter 1970s.

In the early 1970s increased worker participation evolved both through the formation of new structures and the revitalization of old structures. There is reason to believe that unions at the local level became more responsive to their membership, and that they were empowered to assume a more active role in enterprise management. Over 26,000 new union locals were established and local union elections were called in which 87% of the nearly 118,000 officials elected for two-year terms had not previously served.[22] The massive replacements reflected labor dissatisfaction with the policies of the late 1960s, and the identification of incumbent local union officials with those policies.

The shakeup in local union leadership was accompanied by a revitalization of union activity at the local level. Worker assemblies at which issues of material welfare and production were discussed, began to be held regularly again.

Concomitantly, new mechanisms were established to incorporate workers into the enterprise decisionmaking process. Labor gained representation on management boards and a voice in certain aspects of wage policy and production standards. Workers also began to participate in after work meetings at which plan targets, production, health and safety, worker education, and labor discipline were discussed.

As labor became more involved in work-related rulemaking, it came to share responsibility for implementing the policies to which it agreed. Increased worker participation, consequently, contributed to productivity gains. In 1972, for instance, the economywide output per worker jumped 21%, according to Mesa-Lago (1978).[23]

While some analysts have attributed the early 1970s growth in productivity to the use of material incentives, the relationship between the two is tenuous.[24] The government did not even begin to link wages to worker output until 1974 and the process proceeded very slowly thereafter. Moreover, in the first years of the decade there was still considerable excess liquidity held by Cuban households. It is difficult, then, to see a significant connection between the 1972 jump in labor productivity and material incentives. Albeit other factors were at play, it seems likely that the shift in enterprise social relations described above aided in the development of internal incentives (worker motivation) as well as the elicitation of useful ideas for improving the efficiency of the production process.

One other factor that might have contributed to increased worker productivity in the early 1970s was the promulgation of the 1971 compulsory work law. The law specified that workers guilty of absenteeism were to be deprived of vacations, excluded from social benefits (including the work canteen), and, in severe cases, transferred to work camps. The Castro government until then had guaranteed all workers employment and a basic income, to counter the prerevolutionary impact of market vagaries on job insecurity. However, the high rates of absenteeism in the late 1960s demonstrated that labor cooperation could not be assumed. Undoubtedly, the 1971 law, which made work (for men) an obligation in addition to a "socialist right," was implemented in response to the negative experience of the late 1960s. Yet the law may have had a contradictory impact on productivity. On the one hand, one would expect output per unit of labor input to drop as thousands of reluctant and often poorly trained individuals reentered the workforce. On the other hand, in reducing absenteeism, productivity per head (rather than per hour) undoubtedly improved. There is no *a priori* basis for suspecting the net effect of the latter to have been greater than the former.

In the course of the 1970s worker paticipation in enterprise and economywide affairs continued to develop, although the range of activity in which workers had input and their ability to impose changes that management, the Party, or the government opposed, remained limited. Labor, for example, acquired the right to allocate scarce but coveted consumer durables and housing. The central government began to allot work centers, in some instances in accordance with work performance,[25] consumer durables and building materials for housing. The workers then decided how the items would be distributed among themselves — reportedly on the basis of merit (attendance and work record) and need (family size and living conditions).

Workers were increasingly consulted about

national plans as well as their implementation at the enterprise level. According to official reports, the number of workers participating in the discussion and amendment of annual economic plans at the enterprise level increased from 1.26 million in 1975 to 1.45 million in 1980.[26] Workers in 91% of all enterprises discussed the 1980 plan, and their suggestions resulted in changes in the plan's control figures in 59% of all enterprises in 1980.[27] Finally, a 1984 study published by the Cuban International Economic Research Center cites greater worker participation in drawing up the plan for 1984. Around 24,000 proposals were made during the (planning) period, 17,000 of which were approved and included in the plan's figures or given a correct answer in the case of workers' concerns.[28] To be sure, a similar opinion was also expressed by Humberto Pérez, head of JUCEPLAN, who at the National Assembly referred to the popular discussions of the 1984 plan as being the most extensive and significant in the history of Cuban planning.[29]

There is some evidence suggesting that workers felt their participation to be effective and not merely formal. A 1975 study, for example, found that 85% of surveyed workers believed workers must be consulted in enterprise affairs and that 58% felt worker input to be influential.[30] Eighty percent of the 355 workers interviewed in conjunction with another survey, conducted the following year, reported, in turn, that they "always or nearly always" made a personal meaningful intervention at production assemblies.[31] In a 1977 survey conducted by two Cuban researchers of 1,000 randomly selected workers in large enterprises in Havana, the conclusions regarding worker participation in enterprise management were yet more sanguine: there was broad agreement among workers of six different strata that production meetings were held monthly, that there was active worker participation at these meetings, that management was receptive to worker suggestions, and that the deficiencies discussed at these meetings were "always or almost always" addressed and alleviated.[32]

Since 1980 a further impulse to worker participation has been the formation of work brigades (teams of workers consituting a subunit of an enterprise). Brigades elect their own directors, enter into production contracts with their enterprise, organize their own production process and form an accounting unit with its attendant economic incentives. To date the experimental brigades seem to have been successful in both social and economic terms and the intention is to extend their development. By 1985 over 1,000 brigades had been formed in the agricultural sector and over 200 in industry.

In 1975, measures were also taken to decentralize the state apparatus, including the administration of economic activity. The newly formed Organs of Popular Power (OPP) were given responsibility for the administration of locally oriented service, trade, and industrial operations, accounting for 34% of all Cuban enterprises.[33] While local governments are still ultimately responsible to JUCEPLAN (the state planning agency), and while entrenched centralized bureaucracies have not always conceded their sanctioned authority,[34] local OPPs have some latitude in the choice and prioritization of projects and they may name and replace enterprise directors under their control. Local officials, who are popularly elected, must hold periodic meetings at which constituents can discuss matters of concern to them, including the possible recall of incumbents with whom they are dissatisfied. Available information suggests that local meetings are consumed primarily with discussions of bureaucratic deficiencies, including consumer scarcities and complaints about urban services.[35] Accordingly, Popular Power, at the local level, should pressure economic units to perform better; in so doing it should both facilitate domestic savings through more efficient use of existing resources and strengthen "mass" satisfaction with the regime.

Channels have also been instituted to expand worker and mass participation at the provincial and national levels, but the mechanisms are more indirect and more limited in scope, and they seem, to date, to serve more to strengthen "popular" identification with the regime and "popular" responsibility for national decisions than to enable significant input in national decisionmaking. Nonetheless, members of Popular Power at the national level serve on commissions that study economic problems and discuss with JUCEPLAN and the Council of Ministers priorities for the One- and Five-Year Plans.

The second half of the 1970s witnessed Cuba's first Five-Year Plan as well as the gradual introduction of an economic reform known as the "New System of Economic Management and Planning" (SDPE). Each represents the evolving maturity and institutionalization of Cuba's economic organizations.

The SDPE in many respects is modeled after the 1965 Soviet reforms. It attempts (1) to put enterprises on a self-financing basis, (2) to introduce a profitability criterion with its attendent incentives and (3) generally to promote decentralization, organizational coherence, and efficiency. As with the Soviet reform it has met with the obstacles of, *inter alia*, bureaucratic resistance, an irrational price structure, and the weakness of financially-based incentives in a shortage-type economy.[36] Both official studies and privately conducted interviews with enterprise administrators as well as other economic planning personnel who have emigrated to the United States suggest that there are still many snags to be worked out before the SDPE is fully and effectively implemented.[37] To be sure, Cuban officials have repeatedly expressed frustration with the bureaucratic encumbrances of planning as well as with the delays in implementing various parts of the new system. The speeches by former JUCEPLAN President, Humberto Pérez, at the Fourth Plenary evaluation of the SDPE in May 1985 and by Fidel Castro at the recent Party Congress in February 1986 are rather outspoken regarding these ongoing problems.[38]

Since the Cubans began introducing the SDPE they have found this Soviet-styled apparatus too inflexible and too centralized for the Cuban political culture. Each year between 1976 and 1980 innovations, intended to allow for decentralization and greater local initiative, were introduced. By 1980, however, it was decided that the system was being modified too rapidly and this engendered uncertainty about the rules of the game and, in turn, counterproductive instability. A general moratorium on decentralizing reforms ensued and was to last until the Third Party Congress, scheduled for December 1985. Following intensive study, it was projected that major reforms would be carried out to correspond with the Third Five-Year Plan, 1986–90.[39]

Major reports on the deficiencies of the SDPE were presented to the Third Party Congress, but the actual reform proposals were delayed until at least December 1986 when the Congress will be called back into session. In the meantime, the *Grupo Central* and a special planning commission are considering substantial decentralization efforts. As a prelude to such changes, several top planning figures, such as Humberto Pérez and Miguel Figueras, have been replaced. While these personnel changes have been interpreted by some Western scholars as indications that the SDPE has failed, it appears that they are better understood as an attempt to invigorate the planning apparatus with new ideas and fresh relationships unencumbered by previous methods and patterns.

The open, public criticism of the continuing weaknesses of the SDPE is consistent with the longstanding style of self-criticism and high expectations on the part of the Revolution's leadership. It is important to stress, however, that at the same time that the Cubans have expressed dissatisfaction with the slow progress and imperfections of the SDPE they have also repeatedly credited the SDPE with rationalizing and improving the functioning of the Cuban economy. That is, on balance the SDPE has had a positive impact on the Cuban economy. To understand this salutary effect it is necessary to consider several of the concrete, decentralizing measures that have been introduced.

One of the first measures was the 1976 legalization of certain private sector activity. Since then individuals can offer such services as appliance and auto repair work, carpentry, licensed and done on a self-employed basis. The licensing requirement, in principle, enables the government to permit only activities which do not conflict with planned state activity. This economic "opening" does not necessarily represent "capitalist backsliding" but rather represents an effort to correct for the inefficiencies and inadequacies of the bureaucratized state sector and to allow fuller use of labor. License fees also enable the government to profit directly from private activity. In the first month of operation 2,000 people took out licenses to be street peddlers in Havana alone, and in 1981 private contracting cooperatives were responsible for 38% of new housing units.[40] Yet the contained reprivatization of the economy may have the unintended effect of reducing state enterprise productivity if, for example, workers pilfer state enterprises for private supplies, if they absent themselves from work to pursue private jobs, and if they become less committed to their official job because of their preference for the private work. Small-scale private activity is difficult to regulate.

Beginning in the mid-1970s the government also shifted its rural strategy. In raising prices paid by the state procurement agency and by permitting farmers and wage workers to market privately output exceeding their official quotas, material incentives stimulated farm production. The government also pressured farmers less to work on state farms and to cooperate with official agricultural plans. As a result, the number of private farmers who affiliated with state plans dropped — by half, for example, between 1973 and 1977.[41] Instead, the government encouraged farmers to form production cooperatives in which they retained private property rights while benefiting — at least in theory — from economies of scale and a less individuated (and therefore "higher") stage of organization. The number of cooperatives jumped from 44 in 1977 to almost 1,400 in 1985, and cooperatives came to involve over 60% of the land that remained in private hands.[42] Production brigades, along with team and modified household contracting, have also been introduced on state farms in recent years. The agrarian restructuring has had such a positive effect on production that the government has greatly expanded "parallel markets" and begun to tax the income from private sales. (In May 1986, in response to the emergence of a wealthy group of marketing middlemen who transported goods to the peasant markets and as an outgrowth of the expansion of rural production cooperatives and state-run parallel markets, the government announced the demise of the experiment with free farmers' markets. Since the government has raised procurement prices for production in excess of the quota to "free market" levels, peasant income and production incentives should not be significantly affected by this measure.)

The government encouraged one additional domestic resource that previously had been underutilized: women. In 1970 women constituted only 18% of the labor force (the comparable figure for 1958 was 13%).[43] In 1984, the female portion of the labor force rose to approximately 38% — an increase of over 100% from 1970. Unlike in the capitalist countries in the region, in Cuba most working women are employed in jobs that offer employment security, unemployment compensation, pensions, and coverage by the official wage structure. While women are still underrepresented in top managerial and administrative posts, in 1983 53% of Cuba's technical workers were female.[44] Apart from the social progress these figures represent, women's labor force participation reflects an improved use of available human resources. On the other hand, Cuba's birth rate is dangerously low and demographic trends will produce increasing proportions of retired workers constituting a growing fiscal burden for the government. A desire to increase Cuba's birth rate may be one factor behind the absence of new exhortations by the leadership for women to join the work force.

The government also modified its wage and consumer policies. Not only have wages — as previously noted — been increasingly linked to productivity (with slightly less than 50% of all jobs normed by 1979) and enterprise performance, they also have been raised and more closely tied to skill level.[45]

The new wage and labor policies appear to have had a positive economic impact because the government concomitantly modified its consumer policies. There has been a rapid expansion in the number and variety of consumer goods available to the public. Many goods are now available off the ration system. Indeed, while in 1970 95% of consumer spending was on rationed goods, in 1980 this proportion had diminished to approximately 30%.[44] New state housing policies allow dwellers to exchange housing units, tenants to purchase their homes and owners to rent out rooms.

Government concern with consumer preferences is reflected also in the activities of the Institute of Internal Demand. It surveys regularly consumer preferences and makes inventories of available supplies. The Institute relies on a network of thousands of volunteers to provide information on consumer preferences biweekly. The Institute, in addition, publishes a monthly magazine (*Opina*) with consumer information, which since 1976 has included ads for a panapoly of private services available on a fee basis, as discussed below. While consumer allocations entail an immediate investment trade-off, they are central to the government's emphasis, since 1970, on a domestically based accumulation strategy. The increased availability of consumer goods and services should stimulate greater worker motivation and, in turn, productivity.

Although clear progress has been made in using domestic resources more fully and efficiently, the process of decentralization has been slowed by ongoing resource constraints and the related inability of Cuba's planning mechanism to develop slack. Thus, for example, as part of the material incentive scheme of the SDPE, enterprises were to be allowed to retain a certain share of their profits (or reduction in losses) through the so-called "stimulation fund." Largely due to the inability of the Cuban economy to develop slack, however, the "stimulation fund" has been slow to take effect. As long as there are insufficient goods available for enterprises to purchase outside the plan, the ability to retain earnings generally provides only a weak incentive at best. By 1979, only 2.8% of enterprises earned stimulation funds.[47] Moreover, it was originally contemplated by the SDPE that stimulation funds would be used for three purposes: worker bonuses, enterprise; sociocultural expenditures; and, small investments on new machinery and equipment. Yet, again due to the absence of slack and problems with surplus accumulation, the last use of the stimulation fund has been indefinitely suspended.

Although the share of enterprises participating in the stimulation fund system rose to 52.1% in 1985, the absence of sufficient slack in the annual plan and resource constraints significantly weakened the potential motivational impact of the system. A 1984 Cuban study discussed this dilemma (translation theirs):

During the period (1983), the enterprises and agencies of the domestic economy accumulated around 16 million pesos through the application of the material incentives (socio-cultural funds) contained in the SDPE. This situation stemmed from the difficulties encountered in using these monetary resources, since the necessary material counterpart was lacking.

Thus far, the greatest pressure has been in the sphere of construction materials and other investment goods which are in short supply. These (socio-cultural) funds will mainly be used for financing holiday and other celebrations. It is hoped that some resources can be set aside for purchase with these funds in 1984 . . . By the end of 1984, the money accumulated for the purpose should amount to 40 million pesos.[48]

The study goes on to note the intention in the 1984 plan to build in some additional slack through an increase in the state budget surplus (government saving).[49] A major reason for these resource shortages and the inability of the planners to build more slack into the economy, of course, has been the increasing constraints imposed by the external sector.

(b) *External sources of surplus accumulation*

While the government reemphasized a domestically rooted development strategy in the 1970s, it never abandoned its externally rooted strategy. CMEA and Western trade and financing came to be central to the regime's recovery from the political and economic crisis caused by the policies of the late 1960s. As the Western trade gave rise to new problems that have yet to be resolved, relations with the Soviet Union have resurged in economic importance.

Sugar continued to play a key role in the economy. After the crisis caused by the 1970 10-million-ton sugar target, the government lowered the output goal, but successful mechanization of production kept sugar yields high (though below the 1970 record yield) with much less use of labor. Indeed, the government's more permissive stance toward private economic activity, in both the cities and the countryside, undoubtedly partly results from the decline in the need for labor in the sugar sector. Whereas in 1975 25% of cane cutting and 95% of lifting were mechanized, in 1981 46% of cutting and 97.5% of lifting were mechanized, and in 1985 these shares were 62% and 100%, respectively.[50]

The early 1970s success of the sugar strategy, however, is explained less by the reorganization of production or output than by world sugar prices. The world market price of raw sugar reached an all-time high in 1974, and the Soviet Union responded by raising the price it paid for Cuban sugar and by allowing the island to sell on the world market some of the sugar it had contracted to CMEA countries. Island export earnings accordingly reached an unprecedented high.

Although simple correlations between world or Soviet-paid sugar prices and Cuban economic growth do not show a significant relationship, production function estimations (using data from 1961 to 1982) reveal that a 10% increase in the world sugar price is associated with a 0.9% increase in aggregate output per worker in Cuba.[51] High export earnings provide revenue for the importation of inputs that are essential for domestic production and the slack that makes for effective decentralization.

The 1970s export based growth was distinctive in that it was heavily centered around Western trade. Indicative of the trade shift, in 1974 the hard currency earnings enabled the Castro government to import goods and technology from the West for the first time on a significant scale.

The country's capacity to stimulate domestic economic growth through Western trade proved, however, to be contingent on a short-lived favorable conjuncture in the international economy. World market sugar prices dropped from a record high of 68 cents a pound in November 1974 to around seven cents within three years. Since output did not expand at a rate sufficient to offset losses in unit sales, the island's capacity to stimulate capital accumulation through the purchase of imported raw materials, machinery and technology diminished. Moreover, the government was left with outstanding loans contracted when world market sugar prices were high.

Cuba arranged for Western loans to acquire goods and technology unavailable from CMEA. A good portion of the economic improvement of the early 1970s therefore is attributable to the growth in export earnings and the financing to which it gave rise; the previously delineated organizational changes may have been necessary but they in themselves do not fully explain the period's rapid economic growth.

Western banks, in turn, extended the financing to Cuba because their liquidity expanded dramatically following the post-1973 OPEC oil price hikes, and Western governments and banks alike viewed Cuba as a good credit risk given the high world sugar prices at the time and the then excellent credit ratings of the CMEA bloc. Yet, the collapse in the sugar market and then a rise in interest rates caused Cuba's Western debt to climb dramatically, from $660 million in 1974 to

$2.86 billion by the end of 1983.[52] Cuba's debt service (amortization plus interest) to hard currency exports ratio reached 48% in 1980 (or 18.7% in relation to total export earnings).[53] Like other Latin American countries, Cuba has sought and achieved a renegotiation of its debt servicing in recent years; however, because of political discrimination, the terms of the renegotiation have been less favorable than in neighboring countries with more serious debts.

Despite the ongoing burden of its hard currency debt and the continuing drop in world sugar prices, Cuba's debt has not caused a severe dislocation to overall economic activity. As noted in Table 1, Cuban real Gross Social Product grew at about 7% per annum according to official figures during 1981-85 (attributable in large part to improvements in organizational efficiency and the high investment ratios during 1976-80), while the rest of Latin America on average has experienced real national income decreases. (Domínguez, Pérez-López, and others have suggested that the official growth rates for 1981 are unreliable and grossly inflated because they do not adjust for the wholesale price reform — increase — of the same year. Estimates by one of the authors, however, based on constant prices and value added branch weights, suggest a real annual industrial growth rate of 11.7% in 1981 and of 5.5% between 1980 and 1984.)[54] Unemployment in Cuba continues to be negligible at a time when it has reached mammoth proportions in most of Latin America. Moreover, to the extent that labor is unemployed it does not have the same adverse social and economic consequences as in the rest of the region; in Cuba, for example, most unemployed workers are guaranteed a minimum wage as well as free health care and heavily subsidized basic services.

Nonetheless, the burden of debt service has impelled Cuba to concentrate on export production and import reduction. As a result, Cuba has had to curb investment spending. The projected investment ratio (gross investment as a share of Gross Material Product) for 1984 is a relatively low 18%. In contrast, the average investment ratio during 1976-80 was 29.3%.[55] The lower investment ratio, *inter alia*, is likely to slow down the growth rate for the remainder of the decade.

On the brighter side, Cuba has begun to make significant strides in the areas of import substitution and export diversification. Capital goods, consumer durables, chemicals, medicines (Cuba produces 83% of its needs), electronics, computers and steel are all sectors with negligible or zero output in 1958 that are now substantial and growing rapidly. For instance, whereas engineering and capital goods industries (ISIC 37 and 38) accounted for approximately 1.4% of total industrial production in 1959, due to an annual growth rate of 16.6% since 1970, this sector accounted for 13.2% of industrial gross value of output in 1983.[56]

Other products, such as citrus, fish, eggs, nickel, cement and electricity, have expanded several times over since 1958. Cuba currently produces a large share of its new cane harvesters, buses, refrigerators and other durables.[57]

Regarding export diversification, the share of sugar in total exports fell from an average of 86.8% during 1975-79 to 79.9% during 1980-82.[58] This decreased dependence on sugar exports cannot be attributed to lower world sugar prices in the latter period.[59] On the contrary, world sugar prices averaged 11.49 cents per pound during 1975-79 and 18.16 cents per pound during 1980-83. That is, world sugar prices were 58.1% higher on average during the second period, yet the sugar export share was 9.1% lower. Adjusting for sugar prices, then, would indicate even lesser rather than greater dependence on sugar exports. The sugar share in total exports fell further to an average of 75.7% during 1983-84.

From 1976 to 1980 Cuba introduced 115 new export products and in 1981 it introduced 17 more.[60] Whereas non-sugar exports equalled 230.5 million pesos in 1978, in 1982 they equalled 640.9 million pesos.[61] Non-sugar exports grew by an additional 27.5 million pesos in 1983.[62] After falling in 1984, non-sugar exports rose to record levels in 1985.[63] In addition, Cuba earned over 15 million pesos in 1983 from re-exported petroleum; the Soviet Union now allows the island to sell a portion of its oil allotment for hard currency. (These exports, aided by a tripling of domestic crude oil extraction between 1981 and 1984 and a highly successful domestic energy conservation program, accounted for 39.5% of Cuba's hard currency earnings during 1983-85.[64] Indeed, the precipitous drop in oil prices in 1986 is projected to reduce Cuba's 1986 hard currency earnings by over $200 million.)

Yet hard currency commodity export earnings have been insufficient to address Cuba's debt repayment and import needs. Consequently, the leadership has tried still another externally linked surplus accumulation strategy since the late 1970s: direct foreign investment and use of foreign management expertise and technology. In 1977, for example, the Cuban leadership began to encourage contract manufacturing and joint ventures. Contract manufacturing is seen as a strategy whereby Cuba can expand its plant capacity with foreign capital and management skills. Under contract manufacturing arrangements, the government need not draw upon its own capital resources for intial plant investment outlays; rather, the plant is to pay for itself over time through the sales it generates.[65] Joint ventures require the government to furnish some but not all of the capital requirements. The government has implemented joint ventures with Japan in shipping, with Mexico in agricultural machinery-building marketing, with Panama in finance and sugar refining, and it has encouraged a variety of other projects. In 1982, Cuba issued a new foreign investment code, formalizing its efforts to attract foreign investment. The code permits foreigners to own up to 49% of local enterprises, to repatriate profits, and to control labor, pricing and production policies. The code is designed to attract a form of capital that compels foreign investors to absorb more of the risks than is true in the case of finance capital. Negotiations were initiated with a number of Canadian and West European firms but pressure from the Reagan Administration has indefinitely delayed potential investment projects.

Nonetheless, in part because of effective import substitution and export diversification as well as petroleum re-exports, Cuba achieved hard currency trade surpluses in 1982 and 1983 of 749.9 and 450.7 million pesos, repsectively.[66] In this and other indicators Cuba has surpassed the performance targets stipulated in the 1982 Club of Paris debt renegotiations. These gains, it must be stressed, have been made despite the tightening blockade imposed by the Reagan Administration (through effectively pressuring third parties not to trade with Cuba and prohibiting US tourism to Cuba) and the precipitous fall in sugar prices since 1981.

It is also true, of course, that Cuba continues to be cushioned from changes in world market conditions by its privileged position within CMEA. Through significant sugar and nickel price subsidies as well as generous technological transfers and development and payments aid, the Soviet Union has buffered the Cuban economy over the years. This aid, however, has not been nearly as large as claimed by official US sources.[67] Official Western calculations overestimate the sugar subsidy (by comparing Soviet prices with world market prices, rather than with the subsidized prices at which most Western sugar is traded and by using official peso/dollar exchange rates) and they underestimate the costs of the "tied" nature of Soviet aid (i.e., the high cost and poor quality of Soviet imports that must be purchased with the rubles received for Cuban exports). A full assessment of Soviet aid must also take into account the opportunities thereby forgone, namely the consequent loss of aid from and trade with the United States.

Some have suggested that Soviet generosity toward Cuba has diminished in recent years. We, however, have not seen convincing evidence that the 1976 agreement to maintain bilateral terms of trade at unity has been violated. Moreover, it appears that the Cuban debt postponed from 1972 to 1986 has once again been put on hold. Nonetheless, Cuba's slow diversification out of sugar and the continuance of world sugar prices below six cents per pound, along with the ongoing hard currency debt amortization and interest payments, assure that Cuba's foreign sector will be a surplus detractor, imposing constraints on development possibilities through the remainder of this decade.

4. CONCLUSION

This paper has interpreted the evolution of Cuban development strategy in terms of the shifting availability and constraints of domestic and foreign sources of surplus accumulation. It has argued that the operation of these domestic and foreign accumulation sources are intimately linked. We have attempted to demonstrate that by studying the two bases of accumulation our understanding of economic policy options and choices can be enriched.

As Cuba has institutionalized its system of central planning over the last 15 years, it has become apparent to Cuban planners that greater decentralization is needed to mitigate bottlenecks, delays and general inefficiencies as well as to promote the flow of information within the planning apparatus and improve worker/manager motivation. We have described various measures of planning reform with these intended purposes.

The process of successful planning reform and decentralization, however, has been constrained by resource shortages imposed by the external sector since the 1974–75 period of high sugar prices. Since then low sugar prices and costly debt repayment obligations have constricted external accumulation efforts. With the foreign sector a surplus consumer (rather than producer), the necessary resources for meaningful decentralization have been lacking. This circumstance, in turn, threatens to further entangle the planning bureaucracy and to engender frustration in the micro-units of the economy, thereby, hindering the operation of the domestic surplus generation process as well. To be sure, this sequence of reactions represents only a potential force that may make itself felt in varying intensity and may be counteracted by other forces. Notwithstanding the pains of growth and the vagaries of world market pressures, the Cuban economy has emerged with a strong record for the first 25 years of the Revolution. Although short and medium-run hurdles and tests remain, the overall outlook is positive — especially in comparison with Cuba's Latin neighbors.

NOTES

1. Mesa-Lago (1981), pp. 38–39; Brundenius (1981), p. 41. José Luis Rodríguez (1981, p. 127) reports that National Income increased by 19% between January 1959 and August 1960. Also, see Rodríguez et al. (1985).

2. Mesa-Lago (1979). Excellent discussions of the economic management problems as well as the effects of the US blockade are available, among other places, in Boorstein (1968), Gilly (1964), and Fitzgerald (1985).

3. Banco Nacional de Cuba (1982), p. 37.

4. A study by the Cuban Central Bank estimated that Soviet prices on goods purchased by Cuba averaged 50% above world market prices; see Domínguez (1978), p. 156. Also see Radell (1983).

5. Eckstein (1981b), p. 191. Also see Bianchi (1964, pp. 100–161), Forster (1981; 1982), de Llano (1983), and Pollitt (1981).

6. Eckstein (1981a), p. 183.

7. Dirección Central de Estadística (1970, p. 10; 1972, p. 42).

8. Boorstein (1968).

9. Silverman (1973), Roca (1977), Economic Intelligence Unit (1971; 1976).

10. Roca (1977), p. 268.

11. Eckstein (1981b), p. 192.

12. Huberman and Sweezy (1969), p. 118.

13. Calculated from the CIA (1976), p. 5.

14. Brundenius (1981), p. 34.

15. Eckstein (1981a), p. 24. Also see, Centro de Investigaciones de la Economía Mundial (1973), pp. 56–57, 70–85.

16. Mesa-Lago (1978), pp. 25–29.

17. Ibid., p. 38.

18. Mesa-Lago (1981), p. 27.

19. DCE (1972), pp. 30, 34; DCE (1974), pp. 35, 40. The reported increase in the commerce sector is likely to reflect mostly price, rather than real output, increases. Since 1968 the output of the commerce sector has been stated in nominal, as opposed to real, terms in the official statistics. Other sectors continued to be expressed, at least officially, in constant prices.

20. There is an excellent discussion of this and related points in Edquist (1985), Chaps. 4, 6, 7. Also see Benjamin et al. (1984), Chaps. 9 and 12.

21. Two informative studies on the development of the Cuban capital goods industry are: Figueras (1982) and Brundenius (1985).

22. Pérez-Stable (1985), p. 292; Zimbalist (1985), p. 218.

23. Mesa-Lago (1978), p. 39.

24. Ibid., pp. 38–40.

25. Zimbalist and Sherman (1984), p. 376.

26. Fidel Castro, speeches to the 1975 and 1980 Party Congresses.

27. JUCEPLAN (1981). For additional information on worker participation in Cuba, see Harnecker (1979) and Zimbalist (1975).

28. Humberto Pérez, speech before the National Assembly of People's Power on 22 December 1983, reprinted in Granma (1 January 1984).

29. Research Team on the Cuban Economy (1984), p. 36.

30. Pérez-Stable (1976), pp. 31–54. Pérez-Stable's sample of 57 workers was selected non-randomly and disproportionately represented more highly educated and politically conscious workers.

31. Herrera and Rosenkranz (1979), p. 48.

32. Armengol Rios and D'Angelo Hernandez (1977), pp. 156–179.

33. Calculated from Díaz Martínez (1983), p. 81. See also his discussion of ongoing efforts at decentralization, pp. 91–106. Also, see Fuller (1985), pp. 11–13.

34. Roca (1985).

35. See, for one, Harnecker (1979), especially Part Two. Also see Benglesdorf (1985), Chap. 5.

36. JUCEPLAN (1981) and Zimbalist (1985). Among others, see the following Cuban studies: Diaz (1984b), Pérez (1982), Machado (1984), Guzman (1984), Garcia (1984), and de la Rosa *et al.* (1983).

37. The most significant official studies are JUCEPLAN (1981; 1985). Each of these studies is quite candid and blunt about the ongoing problems of inefficiency, bureaucratic delays, etc. They also are clear that the SDPE is yielding increasingly positive returns. Roca (1985) alludes to many of the same problems without placing them in a meaningful perspective.

38. Pérez (1985), JUCEPLAN (1985), Castro (1986).

39. The shortage of trained managerial and technical personnel has been an important constraint on the speed of the decentralization process. This account is based on coversations held in October 1985 and March 1986 with Miguel Figueras, Vice-President of JUCEPLAN until April 1986.

40. Fred Ward (1978), p. 31. Also see Zimbalist and Sherman (1984), p. 381.

41. Domínguez (1980), p. 459. For an interesting treatment of current agricultural policies, see Benjamin *et al.* (1985).

42. *Granma Weekly Review* (27 May 1984), p. 2. Also, see CEPAL (1985), p. 9.

43. Mesa-Lago (1981), p. 118.

44. Research Team on the Cuban Economy (1984), p. 24.

45. Zimbalist (1985), p. 219.

46. Zimbalist and Sherman (1984), p. 383.

47. JUCEPLAN (1985), p. 385. Also see Benavides (1982), Mendez (1984), Codina and Chavez (1981), A. Pérez (1981), and Flores (1984).

48. Research Team on the Cuban Economy (1984), p. 52. A similar point is made in JUCEPLAN (1985).

49. Research Team on the Cuban Economy (1984), p. 61. Also see JUCEPLAN (1985, pp. 42–43) and the closing speech by Humberto Pérez at this *Plenaria* (JUCEPLAN, 1985, p. 45) which analyzes the same problem of insufficiency of material supplies (*contrapartida material*) on which enterprises could spend their retained earnings (stimulation fund).

50. Pérez (1982b), p. 10.

51. Brundenius and Zimbalist (1985), pp. 28–29. See also the ensuing debate in the Spring and Fall 1985 issues of *Comparative Economic Studies*.

52. Banco Nacional de Cuba (1984), p. 18.

53. Eckstein (forthcoming), Table 4.

54. Zimbalist (1986).

55. CEE (various years). Also see Brundenius (1984), pp. 32–33.

56. Calculated from data presented in Brundenius (1985), pp. 11, 20, 21. Also see Figueras (1982) and Brundenius (1985).

57. On the development of an indigenous technological base in Cuba, see Edquist (1985).

58. BNC (1982), pp. 21, 49.

59. Mesa-Lago has claimed that assertions of decreased dependence on sugar are misleading because they overlook the trend of sugar prices. As the numbers in the text demonstrate, Mesa-Lago's observation does not apply to the period since 1975. See Mesa-Lago (1983, p. 34) and Leogrande (1979). Also see Brundenius and Zimbalist (1985) and Brundenius (1985).

60. Humberto Pérez (1982b), p. 39. Also see Rodríguez (1982).

61. BNC (1982), p. 49.

62. BNC (1984), p. 20.

63. BNC (1985a, p. 6; 1985b, pp. 32–35).

64. The results of this conservation effort are most dramatically seen in the sugar sector. In 1976 Cuba's sugar mills consumed 2.11 gallons of oil per metric ton of processed cane; in 1983 the rate had dropped to 0.11 gallons (Research Team on the Cuban Economy, 1984, p. 39). Oil re-exports grew from 96 million pesos in 1980, to 262 million pesos in 1982 and 486 million pesos (8.9% of total exports in 1984). See CEPAL (1985), p. 17) and BNC (1984, p. 31).

65. *New York Times* (25 April 1977), p. 4.

66. BNC (1984), p. 12. A variety of production and price problems, however, turned this balance slightly negative in 1984 and the first three quarters of 1985.

67. CIA (1981), Theriot (1982), Zimbalist (1982). Also see Diaz (1984), Rodríguez (1984), Radell (1983).

68. BNC (1982), p. 13. Also see CECE (1982), Rodríguez (1985), Karlsson (1968, Chap. 17), Acciaris (1984).

REFERENCES

Acciaris, Ricardo, "Cuba's economic relations with Latin America and the Caribbean," *Soviet and East European Foreign Trade*, Vol. 20, No. 3 (Fall 1984).

Armengol Ríos, Alejandro, and Ovidio D'Angelo Hernández, "Aspectos de los procesos de comunicación y participación de los trabajadores en la gestión de las empresas," *Economía y Desarrollo*, Vol. 42 (July/Aug. 1977), pp. 156–179.

Banco Nacional de Cuba (BNC), *Informe Económico* (1982).

BNC, *Informe Económico* (1984).

BNC, *Cuba: Quarterly Economic Report* (September 1985a).

BNC, *Economic Report* (February 1985b).

Benavides, Joaquín, "La ley de la distribución con arreglo al trabajo y la reforma de salarios en Cuba," *Cuba Socialista*, Vol. 2 (March 1982).

Benglesdorf, Carol, "Between vision and reality: Democracy in socialist theory and practice," PhD Dissertation (Cambridge, MA: MIT Department of Political Science, Feb. 1985), Chap. 5.

Benjamin, Medea, *et al.*, *No Free Lunch: Food and Revolution in Cuba Today* (San Francisco: Institute for Food and Development Policy, 1985), Chaps. 9, 12.

Bianchi, Andrés, "Agriculture," in Dudley Seers (Ed.), *Cuba: The Economic and Social Revolution* (Chapel Hill: University of North Carolina Press, 1964).

Boorstein, Edward, *The Economic Transformations of Cuba* (New York: Monthly Review Press, 1968).

Brundenius, Claes, *Economic Growth, Basic Needs and Income Distribution in Revolutionary Cuba* (Lund, Sweden: Research Policy Institute, University of Lund, 1981).

Brundenius, C., *Revolutionary Cuba: The Challenge of Economic Growth with Equity*, (Boulder and London: Westview Press, 1984).

Brundenius, C., "The role of capital goods production in the economic development of Cuba," Paper presented to workshop on "Technology Policies for Development" (Lund, Sweden: Research Policy Institute, University of Lund, 29–31 May 1985).

Brundenius, C., and A. Zimbalist, "Recent studies on Cuban economic growth: A review," *Comparative Economic Studies*, Vol. 28 (Spring 1985a).

Brundenius, C., and Andrew Zimbalist, "Cuban economic growth one more time: A response to imbroglios," *Comparative Economic Studies*, Vol. 28 (Fall 1985b).

Castro, Fidel, "Main report to the Third Congress," *Granma Weekly Review* (16 February, 1986).

Castro, F., Speeches to the 1975 and 1980 Party Congresses.

CECE, "Acciones tomadas por el gobierno de los Estados Unidos en sus relaciones económicas con la República de Cuba," *Economía y Desarrollo*. Vol. 68 (Mayo/Junio 1982).

CEE, *Anuario Estadístico de Cuba* (various years).

CEPAL, *Estudio económico de América Latina y el Caribe. Cuba, 1984* (August 1985).

Central Intelligence Agency (CIA), *The Cuban Economy: A Statistical Review*, 1968–76 (Washington, DC: CIA, 1976).

CIA, *The Cuban Economy: A Statistical Review* (Washington, DC: US Government Printing Office, 1981).

Centro de Investigaciones de la Economía Mundial, *Estudio Acerca de la Eradicación de la Pobreza en Cuba* (La Habana, Sept. 1973).

Codina, Alexis, and Noel Chaviano, "El pago por los fondos productivos en la economía socialista," *Cuestiones de la Economía Planificada*, Vol. 11 (Sept./Oct. 1981).

de la Rosa, Hector, *et al.*, *Los críterios de eficiencia de las inversiones* (La Habana: Editorial de ciéncias sociales, 1983).

de Llano, Eduardo, "La lucha de clases y la segunda ley de reforma agraria," *Cuba Socialista*, Vol. 8 (Sept./Nov. 1983).

Díaz, Julio, "Cuba: industrialización e integración económica socialista," *Economía y Desarrollo*, Vol. 79 (Mar./Abr., 1984a).

Díaz, J., "La aplicación y perfeccionamiento de los mecanismos de dirección en la economía cubana," *Economía y Desarrollo*, Vol. 78 (Enero/Feb., 1984b).

Díaz Martínez, Gilberto, "El Sistema Empresarial Estatal en Cuba," *Cuba Socialista*, Vol. 8 (Sept./Nov., 1983).

Dirección Central de Estadística (DCE), *Compendio Estadístico de Cuba* (La Habana: JUCEPLAN, 1970).

DCE, *Boletín Estadístico de Cuba 1970* (La Habana: JUCEPLAN, 1972).

DCE, *Anuario Estadístico de Cuba 1974* (La Habana, 1974).

Domínguez, Jorge, *Cuba: Order and Revolution* (Cambridge: Harvard University Press, 1978).

Eckstein, Susan, "The debourgeoisement of Cuban cities," in I. L. Horowitz (Ed.), *Cuban Communism*, 4th edition (New Brunswick: Transaction Books, 1981a).

Eckstein, S., "The socialist transformation of Cuban agriculture: Domestic and international constraints," *Social Problems*, Vol. 29 (1981b).

Eckstein, S., "Revolutions and the restructuring of national economies: The Latin American experience," *Comparative Politics* (forthcoming).

Economic Intelligence Unit (EIU), "Cuba, the Dominican Republic, Haiti, and Puerto Rico," *Quarterly Economic Review*, Annual Supplements (1971, 1976).

Edquist, Charles, *Capitalism, Socialism, and Technology: A Comparative Study of Cuba and Jamaica* (London: Zed Books, 1985), Chaps 4, 6, 7.

Figueras, Miguel, *La Producción de Maquinaria y Equipos en Cuba* (La Habana: JUCEPLAN, 1982).

Fitzgerald, Frank, *Politics and Society in Revolutionary Cuba: From the Demise of the Old Middle Class to the Rise of the New Professionals*, Unpublished manuscript (1985).

Flores, Barbara, "Breve Análisis del sistema salarial actual, en los marcos de la Reforma General de

Salarios," *Economía y Desarrollo*, Vol. 78 (Enero/Feb., 1984).

Forster, Nancy, "Cuban agricultural productivity: A comparison of state and private farm sectors," *Cuban Studies*, Vol. 11, No. 2 (July 1981) and Vol. 12, No. 1 (Jan. 1982).

Fuller, Elaine, "Changes expected in Cuba's economy," *Cubatimes* (Jan./Feb., 1985), pp. 11–13.

García, María, "Algunas consideraciones sobre el crédito bancario a corto plazo en la aplicación del SDPE," *Economía y Desarrollo*, Vol. 79 (Mar./Apr. 1984).

Gilly, Adolfo, *Inside the Cuban Revolution* (New York: Monthly Review Press, 1964).

Granma Weekly Review, (27 May, 1984), p. 2.

Guzman, Arturo, "Discurso de clausura del II evento científico de la ANEC del area de ciencias económicas de la u.h.," *Economía y Desarrollo*, Vol. 80 (Mayo/Junio 1984).

Harnecker, Marta, *Cuba: Dictatorship or Democracy?* (Westport, CN: Lawrence Hill and Co., 1979).

Herrera, A., and H. Rosenkranz, "Political consciousness in Cuba," in Griffiths and Griffiths (Eds.), *Cuba: The Second Decade* (London: Britain–Cuba Scientific Liaison Committee, 1979).

Huberman, Leo, and Paul Sweezy, *Socialism in Cuba* (New York: Monthly Review Press, 1969).

JUCEPLAN, *Segunda Plenaria Nacional de Chequeo de la Implantación del SDPE* (La Habana: Ediciones JUCEPLAN, 1981).

JUCEPLAN, *Dictámenes Aprobados en la IV Plenaria Nacional de Chequeo del SDPE* (May 1985).

Karlsson, Gunnar, *Western Economic Warfare, 1947–67* (Stockholm: Almqvist and Wiksell, 1968), Chap. 17.

Leogrande, W., "Cuban dependency: A comparison of pre-revolutionary and post-revolutionary international relations," *Cuban Studies*, Vol. 32 (July 1979), pp. 1–28.

Machado, Jose, "Discurso de Clausura del Activo Nacional del Partido acerca de la rentabilidad de empresas" (Havana, Jan. 1984).

Méndez, Teresita, "Análisis de los fondos de estimulación económica en una empresa industrial cubana," *Ciencias Económicas: Investigaciones*, Vol. 5 (Dic 1984).

Mesa-Lago, Carmelo, *Cuba in the 1970s* (Albuquerque: University of New Mexico Press, 1974).

Mesa-Lago, C., *Cuba in the 1970s*, revised edition (Albuquerque: University of New Mexico Press, 1978).

Mesa-Lago, C., "The economy and international economic relations," in Cole Blasier and C. Mesa-Lago (Eds.), *Cuba in the World* (Pittsburgh: University of Pittsburgh Press, 1979).

Mesa-Lago, C., *The Economy of Socialist Cuba* (Albuquerque: University of New Mexico Press, 1981).

Mesa-Lago, C., "Cuba's centrally planned economy: An equity tradeoff for growth," Paper presented at the Conference on Latin America (Nashville: Vanderbilt University, 3–5 November, 1983).

New York Times (25 April, 1977), p. 4.

Pérez, Aristides, "La prima como forma de estimulación material," *Economía y Desarrollo*, Vol. 80 (Mayo/Junio 1984).

Pérez, Humberto, "Debemos Trazarnos . . ." Speech given at the official closing of III Plenaria Nacional del Chequeo de la Implantación del SDPE (Havana, October 1982a).

Pérez, H., "La Plataforma Programática y el desarrollo económico de Cuba," *Cuba Socialista*, Vol. 3 (June 1982b).

Pérez, H., Speech given to the National Assembly of People's Power (22 Dec., 1983), reprinted in *Granma* (1 Jan. 1984).

Pérez, H., Speech given at the closing of IV Plenaria Nacional del Chequeo de la Implantación del Sistema de Dirección y Planificación de la Economía, to the SDPE, 25 May 1985 (La Habana: JUCEPLAN, 1985).

Pérez-Stable, Marifeli, "Institutionalization and workers' response," *Cuban Studies*, Vol. 6 (1976), pp. 31–54.

Pérez-Stable, M., "Class, organization and conciencia: The Cuban working class After 1970," in S. Halebsky and J. Kirk (Eds.), *Twenty-Five Years of Revolution, 1959–84* (New York: Praeger, 1985).

Pollitt, Brian, "La revolución y el modo de producción en la agricultura cañera de la economía cubana, 1959–81," *Cuestiones de la Economía Planificada*, No. 11 (Sept./Oct. 1981).

Radell, Willard, "Cuban–Soviet sugar trade, 1960–76: How great was the subsidy?" *Journal of Developing Areas*, Vol. 17 (Apr. 1983).

Research Team on the Cuban Economy, *The Most Outstanding Aspects of the Cuban Economy, 1959–83* (Havana: University of Havana, Economic Sciences Area, 1984).

Roca, Sergio, "Cuban economic policy in the 1970s: The trodden paths," in I. L. Horowitz (Ed.), *Cuban Communism*, third edition (New Brunswick: Transaction Books, 1977).

Roca, S., "State enterprises in Cuba under the SDPE," Paper presented at LASA (Albuquerque, April 1985).

Rodríguez, José Luis, "La economía de Cuba socialista," *Economía y Desarrollo*, Vol. 61 (Mar./April 1981).

Rodríguez, J. L., "La economía cubana entre 1976 y 1980: resultados y perspectivas," *Economía y Desarrollo*, Vol. 66 (Enero/Feb. 1982).

Rodríguez, J. L., "Los precios preferenciales en los marcos del CAME: Análisis Preliminar," *Temas de Economía Mundial*, No. 9 (1984).

Rodríguez, J. L., "Un enfoque burgués del sector externo de la economía cubana," *Cuba Socialista*, Vol. 14 (Mar./Abr. 1985).

Rodríguez, J. L., et al., *Cuba: Revolución y Economía 1959–1960* (La Habana: Editorial de ciencias sociales, 1985).

Silverman, Bertram, "Economic organization and social conscience: Some dilemmas of Cuban socialism," in David Barkin and Nita Manitzas (Eds.), *Cuba: The Logic of Revolution* (Andover, MA: Warner Modular Publications, 1973).

Theriot, L., *Cuba Faces the Economic Realities of the 1980s*, Prepared for the Bureau of East–West Trade (Washington, DC: US Department of Commerce, 1982).

Ward, Fred, *Inside Cuba Today* (New York: Crown Publishers, 1978).

Zimbalist, Andrew, "Worker participation in Cuba," *Challenge: The Magazine of Economic Affairs* (Nov./Dec. 1975).

Zimbalist, A., "Soviet aid, US blockade and the Cuban economy," *Comparative Economic Studies*, Vol. 24, No. 4 (Winter 1982).

Zimbalist, A., "Cuban economic planning: Organization and performance," in S. Halebsky and J. Kirk (Eds.), *Twenty-Five Years of Revolution* (New York: Praeger, 1985).

Zimbalist, A., "Cuban industrial growth, 1965–84," *World Development*, Vol. 15, No. 1 (1987).

Zimbalist, A., and H. Sherman, *Comparing Economic Systems: A Political-Economic Approach* (New York: Academic Press, 1984).

·2·

Agricultural Policy and Development in Cuba

JOSÉ LUIZ RODRÍGUEZ

1. INTRODUCTION

There is quite an abundant literature on the Cuban agricultural transformation begun in 1959. Nevertheless, from a factual point of view, the socioeconomic impact of the agrarian reform on the peasantry (*campesino*), has not been sufficiently discussed.

This work was planned to attain a balance that would allow a deep study of the social and economic effects of the above mentioned reform on the Cuban peasantry from 1959 to 1985.

To this end, Sections 2 and 3 offer an introduction which locates the starting point of the discussion and summarizes the character of the agrarian reform undertaken. Section 4 examines the transformations brought about in the land tenure system, agricultural and livestock production results, and the changes in the socioeconomic conditions of the peasants over the last 25 years. The paper finishes with a section aimed at establishing fundamental conclusions to be extracted from the Cuban experience.

Lastly, it must be pointed out that, due to the limitations of this study, the author has included bibliographic notes to enable the interested reader to pursue further study of the subject.

2. THE ECONOMIC AND SOCIAL SITUATION OF THE PEASANTRY PRIOR TO THE AGRARIAN REFORM

(a) *Role of the agricultural and livestock sector in the country's economic development — land tenancy patterns*

Up to 1958 the Cuban economy was largely agricultural, beginning with its development in the colonial period, and continuing in the neocolonial period from 1902 to 1958.

The characteristics of the Cuban economy in the period prior to 1959, can be summarized as follows:

1. It was a structurally deformed and backward economic system due to its primary function as the producer of sugar for the United States-dominated market.[1] During the first half of this century, the Cuban economy entered into a new role in the international division of labor: as a sugar monoproducer and monoexporter, and as a multi-importer of North American merchandise.[2]

2. With such an economic structure, national and foreign investments were assured of cheap and abundant manpower and stable profits. By preventing any agricultural diversification and hindering the country's industrial development, these investments engendered permanent unemployment that fed a vast industrial reserve army.[3]

3. The Cuban economy was highly dependent on the state of the international economy, since reproduction of the system was feasible only through the external sector, thus making this process very vulnerable.[4]

As far as the natural resource endowment for agricultural development in Cuba was concerned, it was estimated that 80% of the land was either cultivated agricultural soil or suitable for cultivation; nevertheless, the actual utilization of land was greatly influenced by the unequal distribution of rainfall, which forced the great majority of cultivation to be carried out between November and April of each year.[5]

With regard to population, estimates in 1960

were that 41% of the total population lived in rural areas, out of which 31% — 860,000 persons — constituted the agricultural labor force.[6]

In the case of land tenure, up to 1958 the ownership structure of the Cuban agricultural and cattle economy was based essentially on *latifundios* or large estates, combined — for certain crops — with *minifundios* or small farms; this condition contributed further to the sector's and country's underdevelopment. As can be seen from Table 1, 9.4% of the owners controlled 73.3% of the land, while 66.1% of the owners held only 7.4% of the land.[7] An overview of land ownership in Cuba and its direct consequences may be obtained by considering the following elements:

revealed by the Cuban agricultural and cattle production results. On the one hand, the essential role of the agricultural and livestock sector in the national economy is reflected by its share in National Income, which fluctuated between approximately 30 and 34% between 1945 and 1954.[9] This share was not an expression of diversified agriculture for if the estimated value of the 1958 agricultural production is analyzed it will be seen that sugar cane constitutes 36.5% of this value; rice 6.2%; coffee 4.4%; and tobacco 6.2%. In total, these four products absorbed 53.3% of the value created in the sector, while the balance of agricultural production only reached 14.2% and livestock production 32.7%.[10] On the other hand, sugar production

Table 1. *Land ownership in Cuba as of 1959* (according to sworn statements by owners affected by the Agrarian Reform Law)

	Surface in hectares	%	Number of farms	%	Number of owners	%
Up to 67.1 hectares	628,673	7.4	28,735	68.3	20,229	66.1
From 67.1 to 402.6 hectares	1,641,440	19.3	9,752	23.2	7,485	24.5
Over 402.6 hectares	6,252,163	73.3	9,602	8.5	2,873	9.4
Total	8,522,276	100.0	42,089	100.0	30,587	100.0

Source: INRA-Legal Department. Data mentioned by Chonchol (1963), p. 75.

— The degree of land concentration, if the sugar latifundio is examined, is still more significant. In fact, the sugar producing companies owned or controlled 2,483,000 hectares of land or 24.6% of the country's agricultural area. Of the total surface controlled by the sugar producing latifundios, 1,800,000 hectares — 17.8% of the total agricultural area — were owned by 13 North American and nine Cuban latifundio entities.

— Only 54% of land controlled by the sugar producing latifundio was planted with sugar cane.

— Considering 1959–60 figures, only 23.6% of the agricultural land was used for planting and of that only 56% was used for sugar cane. Furthermore, only about 10% of the cultivated area was irrigated.

— Finally, according to the 1946 Agricultural Census, 70% of the farm units and 67.6% of total area were operated not by their owners but by administrators, lessees, sublessees, sharecroppers and other occupants (without any deed whatsoever).[8]

The irrationality of this ownership structure is

yields were the lowest worldwide, reaching only 40.3% of those considered to be optimal.[11] Per capita sugar production had diminished from 1.55 metric tons per inhabitant in 1925 to 0.72 in 1958 and Cuba's participation in world sugar production had declined in the same period from 21.8 to 12.6%.[12]

During the 1950s, the Cuban agricultural-based economy was in a state of real crisis. In effect, between 1946 and 1958 per capita sugar cane production grew only at an average annual rate of 1% while all other agricultural production actually decreased at an average annual rate of 1.2%.[13] Nevertheless, according to 1955 figures, 34% of the total imported foods could have been substituted with national products.[14] In summary, it can be affirmed that Cuba was an agricultural country, without agricultural development.

(b) *Socioeconomic situation of the rural population*

In the situation of poverty generated by

underdevelopment, the social stratum which suffered most acutely from its effect was the agricultural workers. In the 1950s this group consisted of 600,000 agricultural workers, some 100,000 sugar workers, and 200,000 peasant (campesino) families of which 140,000 were poor campesinos and semiproletarians.[15]

Under the previously described land ownership conditions, incomes of the Cuban rural population were precarious: taking into account those small farmers forced to pay rent, some 21.8% had an income of less than 200 pesos a year.[16]

The unequal income distribution in the agricultural sector is reflected in figures from the 1945 Agricultural Census. According to these figures, 69.6% of the farms received 27.3% of all incomes generated in the sector, while 7.9% of same received 47.4%. In general, 60.3% of the agricultural and cattle units received annual incomes of less than 1,000 pesos; 27.0% received between 1,000 and 2,999 pesos and 12.7% received 3,000 pesos or more.[17]

Nonetheless, other studies made at the end of the 1950s revealed that 50.6%-plus of all peasant families sampled had an income of less than 500 pesos per year; 42.1% had incomes between 500 and 1,000 pesos; and 7.2% earned 1,000 to 1,200 pesos in a year. Average yearly family income for sampled peasants was 548.7 pesos, including income imputed from crops grown for their own consumption. Average budgetary expenditures were as follows: 69.3% for food; 14.1% for clothing; 7.5% for various services; 1.7% for housing and 7.4% for miscellaneous expenses.[18]

In general, agricultural workers' level of income is estimated to have been 25% of the average national income per inhabitant in 1956. The standard of living of the rural population at the end of the 1950s can be summarized as follows:

— The food of the rural population consisted basically of rice, beans and *viandas* (plantains, cassava, bananas). Only 11.2% of the rural population drank milk; 4.0% ate meat; 3.4% bread; 2.1% eggs; and less than 1% ate fish.

— As for health conditions and medical care, it is estimated that 14% suffered or had suffered from tuberculosis; 13% had had typhoid fever; and 36% intestinal parasites. Free governmental medical attention had been received by only 8% and there was only one rural hospital in the entire country.

— Forty-three percent of the rural workers were illiterate while 44% had never gone to a school.

— Finally, with regard to housing, according to the 1953 Census figures, 74.0% of all rural dwellings were in poor condition — 1956 figures show that 60.3% of agricultural workers' dwellings had wooden walls, roofs of *guano* (palm tree leaves) and dirt floors; 64% had no latrine or WC; 82.6% had no bath or shower; and only 7.3% had electricity.[19]

3. THE CUBAN AGRARIAN REFORM

(a) *The socioeconomic development strategy of the Cuban Revolution and the agricultural sector*

From its inception the Cuban Revolution encompassed an integral development concept which included both economic and social aspects. In respect to this, in 1953 Castro stated: "The problem of the land; the problem of industrialization; the problem of housing; the problem of unemployment; the problem of education and the problem of the health of the people; these are the six points, the solution to which we shall direct all of our efforts, along with the acquisition of public liberties and political democracy."[20]

In his speech, "History will absolve me," Castro presented a program of measures which, on the one hand, would lead to overcoming two fundamental obstacles for the country's economic development: its agriculturally-based, deformed economic structure and its dependency relations with the United States. On the other hand, priority attention to serious social problems would offer the complement in terms of indispensable social development to fulfill this program of basic transformations.

After 1959, the implemented economic policy was essentially directed towards structural transformations which were an indispensable prerequisite to the economic development of the country, with the agricultural and cattle sector playing a fundamental role. Along with the above, measures would be implemented to redistribute income in favor of the poorest; to diversify agricultural production; to industrialize the country; and to reorient international economic relations.[21]

Among the basic structural transformations undertaken during 1959–60, the 17 May 1959 Agrarian Reform Law was essential. This law began with the abolition of latifundios and had as social objectives reserving for the state economy all those lands which were not under cultivation and guaranteeing property rights for the small and medium size farmers who worked the land without owning it. From the economic point of view, the idea was to create an internal market needed by industry, by raising the standard of living in the countryside.[23] The Law of Agrarian

Reform ". . . was not conceived only as a structural change for agrarian property, but also as a radical transformation of the agricultural structure with the objectives of ending the sugar monoculture, creating a national food strategy through import substitutions; supplying raw materials necessary for industry and expanding those agricultural exports."[24]

The role of the agricultural sector in the country's economic development program would subsequently be visualized as an indispensable element. Nevertheless, its specific position among the fundamental priorities of the adopted economic development strategy would be modified over time. Thus, in the period 1961 through 1963, in the context of an accelerated industrialization process, crop development depended upon the possibility of moving from an extensive agriculture to a low cost intensive one, assuming the presence of favorable soil and climate conditions.[25]

In 1964 the development strategy was reformulated. The new development strategy objectives basically focused on creating the conditions for Cuba's economic transformation through accumulation. This strategy — that would be applied between 1964 and 1975 — was based on the utilization of the sugar cane monoproducing and monoexporting structure in order to create a technical/material base for the country's further industrialization by technically re-equipping agricultural production. Studies undertaken revealed that the bases upon which Cuban development would depend were sugar, cattle and food production.[26]

The creation of conditions necessary to move towards industrialization culminated in 1975. As of 1976, industrialization itself became the core of the economic development strategy. Nevertheless, this program would continue to rely significantly on the agricultural sector as an export funds producer and as generator of an import substitution process, as well as an essential supplier of food for the people and raw materials for industry.[27] In general, it can be said that in the years since 1976 the agricultural sector has continued to play an essential role as the basis of an industrialization process that is still in its early stages in Cuba.

(b) *Characteristics of the agrarian reform laws*

The need for a change in the Cuban land ownership system was recognized in article 90 of the 1940 Constitution, which abolished the latifundio. However, the complementary legislation to attain the objectives mentioned in article 90 was never approved.

The first legislative measure to this end was adopted on 10 October 1958. Law No. 3 was approved by the Rebel Army fighting against Fulgencio Batista's government in the Sierra Maestra. The objective of the law was to convert the peasants into allies of the Revolution through the free handing over of up to 26 hectares of land to those who owned land or could make use of an extension of up to 67 hectares — assuming they were already working it. The law also mentioned article 90, but the latifundio problem was postponed for the future.[28]

After 1 January 1959, an agrarian reform bill was introduced which, though citing Law No. 3 as its antecedent, would surpass it by far. This bill would become the Agrarian Reform Law on 17 May 1959.

As previously mentioned, this law was aimed at the attainment of significant socioeconomic objectives. This is evident from the law's fundamental features:

— The law outlawed the latifundio, by limiting the maximum extension of land possessed by each owner to 402.6 hectares, though in some cases it was extended up to a maximum of 1,342 hectares.

— In the future, land could be owned or inherited only by Cuban citizens or entities.

— A 26.8 hectare vital minimum was established to be given free to all of those working the land; at the same time it was made possible for beneficiaries to acquire up to 67.1 hectares.

— The law anticipated a compensatory payment to expropriated owners taking into account the value of their properties as declared for tax payment purposes before 10 October 1958. Such compensation would be payable in bonds of the Republic of Cuba for a 20-year term at an interest rate not higher than 4.5%.

— Properties received free as a result of the law could be transferred only by heredity, sale to the State or an authorized alternative. Similarly, rental contracts, sharecropping, usufruct property or mortgages were banned.

— Agricultural cooperatives were encouraged whenever possible.

— An independent governmental organization was established for the implementation of the law, namely, the National Institute for Agrarian Reform (INRA). It dealt not only with tasks related to land redistribution and agricultural production, but also education, health and housing in the rural areas.[29]

This 1959 Agrarian Reform Law contained some features which should be highlighted:

— The land redistribution process was not

limited to a mere colonization of government owned virgin land, it also eliminated the latifundio which constituted the greatest obstacle to the development of agriculture and the national economy as a whole.

— Flexible solutions were adopted in the land redistribution process. On the one hand, land was given to all non-owner peasants. On the other hand, there was no land redistribution in the great sugar producing and agricultural latifundios out of consideration for the interests and objectives of their agricultural workers.

Thus, for the sugar producing latifundios, an advanced intermediate form of social property was established by creating the sugar cane cooperatives; for the cattle latifundios a direct governmental property system was implanted.

Through this solution unnecessary parcelling of large agricultural plantations was avoided. This measure represented undoubted advantages from the technical-production point of view, while at the same time establishing social relations more in accordance with the interests of society as a whole. This solution contained clear advantages from a political point of view. A significant transformation resulted from the 1959 Law for Agrarian Reform.[30]

Table 2 compares a sample encompassing 84.6% of the available agricultural area before the agrarian reform with another sample covering 74% of available agricultural land in August 1961.

Changes brought about by the Agrarian Reform Law may be summarized as follows:

— Latifundios were eliminated as a distinctive feature for the land ownership system in Cuba.

— Social property became dominant in the land ownership structure, with the small campesino groups following in importance.

— Even though large and medium size private agricultural plantations were not eliminated, their notorious influence was limited within the sector.[32]

It must be mentioned, when referring to the redistributive effects of the agrarian reform, that its direct consequence was the transfer of some 40% of the land to public control between May 1959 and May 1961. Other laws, with the same objectives, were applied by the revolutionary government during the same period.[33]

In summary, immediate consequences of the May 1959 Agrarian Reform Law were: Latifundio properties were eliminated and some 60% of the land was redistributed to small campesinos and the government; the government agricultural sector was created to control some 40% of the properties; campesinos were liberated from lease payments, intermediaries' and moneylenders' exploitation; superior agricultural production forms were created; and, there was a significant income redistribution in favor of lower income groups, enhancing the Cuban internal market.[34] Still, problems related to the Cuban land ownership system were not totally solved with the 1959 law.

In effect, even with the preponderance of the governmental sector and small campesinos, agricultural farming in lots over 67.1 hectares that were privately owned covered 2,102,860 hectares in August 1961 (see Table 2). By mid-1963, such land accounted for 1,717,500 hectares and was distributed as shown in Table 3.

Table 2. *Changes in agrarian property*[31]

Property size	Before the Law for Agrarian Reform (May 1959) Number of properties	Size in hectares	%	After the Law for Agrarian Reform (August 1961) Number of properties	Size in hectares	%
Private property up to 67 hectares	28,735	628,673	7.4	154,703	2,348,151	31.5
Private property 67.1 to 402.6 hectares	9,752	1,641,440	19.3	10,623	1,725,404	23.2
Private property more than 402.6 hectares	3,602	6,252,163	73.3	592	377,456	5.1
Social property	—	—	—	—	2,995,549	40.2
Total	42,089	8,522,276	100.0	—	7,446,560	100.0

Table 3. *Medium and large owner distribution in Cuban agriculture around mid-1963*[35]

Farm size	Number of owners	Size in hectares
67 to 134 hectares	6,000	607,500
134 to 268 hectares	3,000	610,000
Over 268 hectares	1,500	500,000
Total	10,500	1,717,500

Though this figure was only 17% of the total agricultural area, these lands produced over one million tons of sugar, while important grazing areas were controlled for cattle development.[36] Even so, from May 1961 to September 1963, these owners hostilely confronted the development program, hindering wherever possible the agricultural production advances because they were conceived within a socialist framework.

Due to opposition from remaining large landholders, the Cuban government was forced to adopt appropriate countermeasures, embodied in the second Law For Agrarian Reform of 3 October 1963. This provided for the expropriation of land from those who owned more than 67 hectares. This law also included indemnification procedures by fixing a 15 peso monthly payment for each 13.4 hectares expropriated. These payments covered a 10-year period with minimum and maximum limits set between 100 and 250 pesos, respectively.

In this way the government came to control some 5,098,409 hectares, or 68.5% of all recorded land according to the 1961 Cattle Census (Table 2), or 76.6% of all of the agricultural land in Cuba according to other estimates.[37] This law created the necessary conditions for the planned development of the agricultural sector by completing fundamental changes in the agrarian property structure.

4. LAND OWNERSHIP SYSTEM AND AGRICULTURAL DEVELOPMENT IN CUBA

After the second Agrarian Reform Law, state property was predominant within the Cuban agricultural sector. However, private agricultural production would significantly stand out in certain crops, particularly in those needing intensive attention and specialized cultivation. Organizational forms adopted by state agricultural production as of 1959 were varied and changed according to differing historical experiences.

Initially, the countrywide organization in charge of agrarian policy implementation was the National Institute for Agrarian Reform (INRA). INRA continued in this capacity from 1959 to 1976. As of 1976, this function was taken over by the Ministry of Agriculture and, partially, by the Ministry of the Sugar Producing Industry.

From the product composition point of view, non-sugar cane agriculture tended towards a base level diversification for the 1963–64 period. At that time a regional specialization phase would begin to function at different stages.

During the 1966–75 period plans were developed to integrate regionally all of the land — including the different forms of private property — under the same specialization pattern. This occurred, however, in a more flexible manner. During this period, horizontal integration of the sugar sector was implemented through agro-industrial complexes, in which sugar producing agriculture and manufacturing were joined under the same entity.

Prior to 1959 the campesino private sector was organized into different associations, which were controlled by great estate owners. After the May 1959 Agrarian Reform Law was enacted, these estates were under obligation to form associations grouped in accordance with their land ownership level. Thus, the National Association for Small Producers (ANAP) was formed in May 1961, and the vast majority of farmers with up to 67 hectares of land were grouped under it. This organization was constituted to defend campesino interests, and until 1963 undertook administrative functions as well as sociopolitical tasks. Under the umbrella of this association the Cuban peasantry would develop diverse forms of collaboration and cooperation. These activities developed on two different levels: within the private sector itself, and under different means of association with the state sector. In the private sector, the campesino association[38] and credit and service cooperatives[39] evolved. The latter may be considered as an incipient form of production cooperatives whose tendency to grow persisted from 1959 until the end of the 1970s, when agricultural production cooperatives were started. In the category of state property, the agricultural societies,[40] arose as an initial means of production cooperation among the peasantry. This form, however, was not fully developed (as cooperatives for agricultural production) until 1977.[41]

During the 1966–75 period it was felt that the conditions existed for integrating the majority of private campesinos into the system of state property. This process occurred in two ways: by developing special plans and by developing integral plans.[42] With these plans it was possible to integrate 753,532 hectares of land from the private sector into the different forms of state property by late 1972.[43] This figure represented approximately 31% of the land possessed by farmers in 1971.[44] Nonetheless, this form of integration into state production did not produce the expected results. In essence, the forms of state property adopted were costly and lacked the necessary material incentives to convert the small campesino into a more efficient producer.[45] Conversely, more recently, the development of agricultural producer cooperatives has proven totally advantageous. Legalized by the 1982 Law for Agricultural Cooperatives, producer cooperatives numbered 1,378 by 1985, each with an average size of 792 hectares. Thus, cooperative land represented 61.3% of the total land owned by private campesinos.[46] In 1983 81.5% of all profitable production cooperatives had an average cost of 0.7 cents for each peso of production — showing a good level of profitability.[47]

By 1985 the structure of agrarian property in Cuba was comprised of 80% government property, 12% cooperative property and 8% individual private property.[48]

(a) *Economic achievements in the agricultural sector*

The post-1959 changes in the Cuban land ownership system would create the favorable conditions for the country's agricultural development. Such conditions alone, however, would not have ensured development; indispensable resources had to be supplied. For instance, under conditions prevalent in Cuba in 1959, it was necessary to use resources to build up the infrastructure. Installing irrigation systems, and building roads and appropriate facilities for cattle development were necessary prerequisites for the better exploitation of land and livestock. It was also necessary to increase areas under cultivation by incorporating new lands. Further, it was necessary to initiate a slow process of mechanizing the agriculture sector which had been operating from a very rudimentary base and had low productivity levels. Finally, it was necessary to make large human capital investments so that agricultural manpower could acquire the requisite skill levels to work under the new conditions. It was due to these circumstances that the investment process was far more prolonged than originally planned.

The government invested 11,179.8 million pesos in agriculture from 1960 to 1985, a sum representing 26.1% of total investments in the country during this period. These investments resulted in the following improvements over the situation in 1959:

— A 16.5% increase in the country's cultivated area;
— an eight-fold increase in the stock of tractors;
— a ten-fold increase in the application of fertilizers;
— a four-fold increase in the application of pesticides;
— a 125-fold greater capacity to dam water, with an irrigation area covering around one million hectares; and
— construction of around 3,000 new agricultural and industrial facilities in rural areas.[49]

With regard to the government sector, campesinos have been granted 1,274 million pesos in credits for the 1961–84 period. From 1961–67 private farmers were granted credit at a 4% interest rate for loans under 5,000 pesos and 6% for larger loans, while cooperatives were charged 3% for loans under 5,000 pesos and 3.5% for larger loans.[50] Between 1967 and 1976 small farmers were exempt from tax and interest. It must also be noted that from roughly 1967 to 1974 Cuban private farmers who were incorporated into governmental production plans received free inputs, manpower and technical services.

Since 1977 the Cuban government has granted considerable aid for agricultural cooperative development. Thus, by 1984 agricultural producer cooperatives had a capital stock which included 393 cane cutter combines, 5,115 tractors, 1,007 trucks, 518 cane lifters, 4,573 long arrow carts and 1,177 irrigation systems.[51] Special reference must be made to the government policy (especially since 1976) of raising purchase prices, which has stimulated small peasant production.[52]

Finally, during the last few years several supplementary marketing mechanisms have been promoted for the above-plan production of cooperatives and small farmers. Thus, in 1980 free peasant markets were approved for such sales at prices determined by supply and demand.[53]

Hence, Cuban agrarian policy since 1959 has ensured the realization of the production potential inherent in every progressive change of land ownership systems. These policy results can be observed in agriculture production statistics. Gross production in the state agricultural sector grew at a 2.8% average annual rate between 1962 and 1985, while the value of agricultural services increased yearly by an average of 6.1% (see

Appendix Table A2). Sugar cane production between 1962 and 1984 increased at an average rate of 3.5%. State sugar production grew at a yearly average rate of 6.9%, while private sugar production suffered a yearly decrease of 2.0%, reflecting the change in the land ownership system during that period.

Nevertheless, if yields are examined the important role of sugar cane production from private sources can be observed. Overall yields per hectare increased at an annual average rate of 2.7%, with a 2.3% rate for the state sector and 2.9% for the non-state sector (see Appendix Table A3).[54]

There were significant advances in the production of fruit and vegetable crops between 1962 and 1984; the crops enjoyed an annual average yield increase of 4.8% and 3.3% respectively. Edible tubers and roots grew at 2.5% a year, while cereals grew at 2.6%. In the majority of the cases, with the exception of beans, vegetables, cacao beans and tobacco the state sector is the principal producer (see Appendix Table A4). With regard to livestock products the most significant advances have been in egg production, which had a 12.5% annual growth rate from 1962 to 1984. Annual pork output increased at 12.8%, milk 6.3%, and fowl 5.4%. In all of these cases, the main producer has been the state farm (see Appendix Table A5).

In general, the faster growth of the state sector relative to the non-state sector (private campesinos and cooperatives) is more evident from 1975 to 1984 (see Appendix Table A7). In addition, it must be noted that, according to different evidence, the real production cost for the private sector is very low since several other factors, such as family work, were not computed.[55]

It must be noted that agricultural development has encountered a variety of difficulties related to the initial weaknesses in the sector, the complexities inherent to the process and errors in economic policy particularly during the early years. Still, the results have been positive and have tended towards higher yields.

(b) *Changes in the socioeconomic situation since enactment of the Agrarian Reform Laws*

With the advances attained in Cuban agriculture and livestock production have come substantial improvements in the life of the rural population since the development concept was first applied in Cuba in 1959. The determining elements in the socioeconomic status of the Cuban rural population include: employment, salaries and incomes; social security and assistance; education; health; and housing.[56] A brief look at each reveals the improved standard of living in rural Cuba.

(i) *Employment, salaries and incomes*

After the enactment of the Agrarian Reform Laws agricultural unemployment disappeared. Employment increased in the agriculture sector by 53% between 1958 and 1962, from 598,000 to 995,000 people. Later there would be a chronic manpower shortage in the rural zones.[57]

With regard to incomes, 85% of the campesinos stopped paying rent (estimated at 9.6 million pesos a year). With the creation of the *Tiendas del Pueblo* (People's Stores) and their price policy the purchasing power of the rural population increased by more than 60%. The global income redistribution in favor of the campesinos between 1959 and 1962 can be estimated at between 250 and 300 million pesos, representing between 9 and 11% of the national income for that period.

Thus, from an average per capita agricultural income estimated at 91.3 pesos a year in 1956,[58] a sample obtained between 1975 and 1976 put annual per capita incomes for peasant families at between 719.3 and 840.7 pesos and for non-peasant rural families at between 597.4 and 712.4 pesos.[59] Between 1966 and 1984 the average salary paid in the state agriculture sector (see Appendix Table A6) increased at an annual average rate of 3.6%.

(ii) *Social security and assistance*[60]

Before 1959, the majority of the 50 social security institutions in Cuba did not include agricultural laborers. Social security coverage for state agricultural workers was introduced with the Social Security Laws Nos. 1100 in 1963 and 24 in 1980. Out of total social security benefits granted between 1972 and 1981, 28.7% were for agricultural workers. Private farmers would benefit through the INRA's special resolutions (266/1961, 120/1966, 178/1967) covering 34,059 persons in 1981. In 1973 private farmers were covered by Law 1259, and finally the 1983 Decree — Law 65 — put into operation a social security system for salaried workers. Today, the Cuban social security system covers practically 100% of the rural population.

(iii) *Education*

Before 1959 the number of schoolrooms in the fields could not hold more than 35% of the student population, illiteracy before 1961 was 41.7% and it was estimated that more than 44% of the rural population had never been to a school. In 1953 the level of educational attainment for the rural population was as follows: 96% of the population had less than six years of primary education, 4% had just six years, and 0.5% had nine or more years.[61]

After 1959 illiteracy was reduced to 3.9% of the population and after 1981 to 1.9%; in 1959 alone, more than 10,000 new schoolrooms were created nationwide and the number of teachers tripled. Special plans were developed to take education to the mountainous areas which, between 1959 and 1961, received 3,000 volunteer teachers and were provided with 150,000 scholarships for campesino girls to ensure their studies. School enrollment in rural areas, which in 1961 was lower than 40.0%, reached 88.0% in 1981 for children between the ages of 6 and 16.[62] In 1981 the educational attainment for the rural population six years of age and older was: 51.9% had less than six years of primary education, 48.1% had six or more years, and 13.9% had nine or more years.[63]

(iv) *Health*

At the end of the 1950s Cuba had only one rural hospital, government medical attention was received by only 8% of the rural population, and it was estimated that 14% of all rural workers suffered from tuberculosis and 13% from typhoid fever.

Between 1959 and 1983, 55 rural hospitals and 218 medical posts were built in agricultural areas.[64] By 1981 medicine provided for free by the government reached 100% of the rural population; diseases such as tuberculosis, typhoid fever, diptheria and polio have practically disappeared from rural areas in Cuba.

(v) *Housing*

According to 1953 data, 74.0% of all rural houses were in poor condition; 23.0% were considered to be in more or less fair condition, and 3.0% were in good condition. In 1956 it was reported that 63.96% of the dwellings of agricultural workers had neither a WC nor a latrine; 82.62% had no bath or shower, and only 7.26% had electricity.

In 1981, 20% of all rural dwelling were in poor condition; 65% were in more or less fair condition and 15% were in good condition. Of these dwellings 79.7% had sanitary facilities, 15.5% had bathrooms with showers, and 45.7% had electricity.[65] Besides this, campesino life has been modified by the gradual elimination of isolated dwellings, and the development of communities that have all the sociocultural advantages of modern urban life.

5. CONCLUSIONS

The agrarian reform has constituted an essential part of the process of major economic, political and social transformation in Cuba since 1959. This has ensured that changes in land ownership forms would find the necessary continuity to produce results. The social and economic policy formed in the Cuban agricultural sector during the last 26 years has fostered positive results in every respect. Flexibility and adaptability in policy formation, appropriate to the diverse land ownership forms during different periods, have underwritten this success.

NOTES

1. Between 1902 and 1958 some 72% of all Cuban exports went to the United States, while 71% of all imports were from the US. Computations are based on the work of Zanetti (1975), Tables 8 and 9.

2. Between 1902 and 1958, 82% of all exports were from sugar and 10% from tobacco. See Zanetti (1975).

3. It has been estimated for 1957 that unemployment was 16.4% for the economically active population (EAP) to which was added another 17.7% who were underemployed. In general, an average 33.5% of the country's EAP was unemployed or underemployed. See CIEM (1983), pp. 14–15.

4. The ratio of imports plus exports to National Income experienced no significant change during the neocolonial period, moving from 73% in 1903 to 70.6% in 1958.

5. During the dry months (November to April) rainfall decreased to 300–400 millimeters while during the rainy months (May to October) it increased to 1,000–1,100 millimeters. See Chonchol (1963), p. 71.

6. Chonchol (1963), p. 73.

7. *Ibid.*, p. 72.

8. *Ibid.*, pp. 72 and 76.

9. See "Agricultura y ingreso nacional" (1955), p. 592.

10. Information from Departamento de Producción, INRA, cited by Chonchol (1963), p. 78.

11. The percentage was 66.6% for rice and 24.9% for coffee. Figures based on information by Pino Santos (1960), p. 74.

12. Computations based on information from *Anuario Azucarero de Cuba* (1963) and CEDEM (1976).

13. Computations based on information by Brundenius (1984), p. 18.

14. Calculations based on information from "Agricultura y desarrollo económico" (1956), p. 630 and Zanetti (1975), Chart 6.

15. See C. R. Rodríguez (1983b), pp. 307–318.

16. CIEM (1983), p. 33.

17. Computations based on data mentioned by Acosta (1972a), p. 73.

18. Information from the 1956 survey by Agrupación Católica Universitaria of a sample formed by agricultural worker families published in *Por qué Reforma Agraria* and cited in Chonchol (1963), p. 88. It is considered that this survey suffered from serious methodological deficiencies that caused it to undervalue negative aspects of rural life. See Pollitt (1967).

19. Based on data from research by Agrupación Católica Universitaria mentioned by Chonchol (1963), pp. 87–88 and CIEM (1983), pp. 25 and 33.

20. Castro (1973), p. 43.

21. See J. L. Rodríguez (1979).

22. See C. R. Rodríguez (1983a), p. 213.

23. See *Primer Forum Nacional* (1959), p. 17.

24. See Kalecki (1960).

25. See "La producción agrícola y animal en Cuba" (n.d.) and Figueras (1965), pp. 20–29.

26. See *Plataforma Programática del PCC* (1976); and *Lineamientos económicos y sociales* (1981), pp. 115–132.

27. In this respect see Martínez (1978) and Acosta (1972).

28. See Chonchol (1963), pp. 89–94 and Acosta (1972b), pp. 93–105.

29. With regard to available figures on these changes, it should be noted that they have certain deficiencies in method, for instance, they do not always consider the total agricultural area of the country and information is sometimes for different time periods. Nevertheless, fundamental trends are clearly expressed in all cases.

30. See Chonchol (1963), p. 75 and C. R. Rodríguez (1983a), p. 218.

31. According to other estimates concerning the effects of the Agrarian Reforms up to July 1963, a total of 6,073,000 hectares of land were expropriated, corresponding to 60.3% of all agricultural land in the country; 3,768,000 hectares, or 37.4%, were passed to the state sector, while 883,000 or 8.9% were given to small campesinos with farms as large as 67.1 hectares. Thus it is estimated that land ownership structure, to mid-1963, was:

State sector land	3,768,000 hectares,	37.4%
Land owned by small farmers	2,348,000 hectares,	23.3%
Land owned by large and medium farmers	3,954,000 hectares,	39.3%

From all points of view these figures seem far too high. See McEwan (1981), pp. 45–46.

32. See J. L. Rodríguez (1979), p. 125.

33. See C. R. Rodríguez (1983c), p. 248.

34. See "Una evaluación de la reforma agraria en Cuba" (1972), p. 179.

35. See note 32.

36. On these see McEwan (1981), Chap. 10, and Acosta (1972b), pp. 105–109.

37. On these aspects related to the government sector see C. R. Rodríguez (1983a); Bondarchuk (1980); Yañez (1981); and del Monte (1981).

38. ". . . Represent small campesino voluntary unity with a politically fundamental purpose (. . .) In the campesino associations farmers do not relinquish any of their rights as owners (. . .) Nevertheless, there are cases where, for a better performance in their tasks, they obtain some equipment unobtainable for any of the members as private individuals (. . .) Under these conditions such equipment would become association owned and would not belong to any given campesino." Also, within these associations several basic forms of cooperation were developed with regard to planting or harvesting operations (Self Help Brigades or FMC-ANAP Brigades). See Acosta (1973a), pp. 152–153. Between 1979 and 1980 the majority of campesino associations became credit and service cooperatives. By 1984, 91% of small campesinos were integrated into these types of cooperatives. See Ramirez (1984), p. 15.

39. The fundamental purpose of these cooperatives was related to obtaining credits and contracting several productive services, for which the cooperatives became a legal entity. Simultaneously the cooperative establishes its production plans as a precondition to obtaining credits and services in such a way as to enter into a contractual commitment with the government. Campesinos may choose to integrate part of their means of production, including the land, and leave some of them unintegrated. Acosta (1973a), p. 154.

40. "Agricultural societies are formed based on family groups that voluntarily have decided to join their

lands, equipment, work animals, etc., transforming the small individual properties into a collective property, where work is performed in accordance with rules established by them, and incomes are proportionally shared, according to work done by each group member." *Ibid.*

41. For an analysis of initial difficulties experienced by Cuban agriculture see C. R. Rodríguez (1983a), pp. 224 and 227.

42. ". . . The *special plan* is a territorial production group formed by private and independent government farms working under a special government production plan for agricultural lands. Material and technical supplies, as well as harvesting and storing of products are all under government guidance. The *guided plan* is a variant of the above in which private and government farms work, at their own expense and independently from one another, in a production corresponding to the special plan for the given agricultural zone . . . The *integral plan* represents a territorial production enterprise in which land and other means for production of small farmers are integrated in a government farm. Private owners receive wages for work done and rent for their land used by the *integral plan*. See Bondarchuk (1980), p. 158.

43. DOR del CC del PCC (1975), p. 42.

44. Computations estimated according to information from ANAP cited by Acosta (1973a), p. 158.

45. The small campesino integrating himself into government plans received supplies and manpower as support and at no cost; this ensured significant profits independent of production results. There were cases in which campesinos charged 1,000 pesos for each 13.4 hectares of land given in usufruct to the government. This guaranteed high incomes and no responsibility for the results of the use of the land. In this respect see Martin (1982), pp. 17–73, 80–81 and 104–107.

46. Castro (1986), p. 76.

47. Ramírez (1984), pp. 8 and 11.

48. Author's estimate.

49. See Castro (1984).

50. Reimbursement periods for these credits were one year for production loans and five years for development loans. Besides this, there was a policy forgiving the repayment of credits to campesinos affected by natural disasters. See Acosta (1972b), pp. 110–115.

51. Ramírez (1984), p. 10.

52. This policy was notably improved after 1976.

53. See Decreto 66/1980, Comité Ejecutivo del Consejo de Ministros (1980). In the free campesino markets, products were also sold by owners of small lots and also by non-campesinos. These small properties covered an area estimated at some 247,700 hectares in 1981. See Figueroa and Garcia de la Torres (1984), p. 39.

54. The causes for better production in the non-government sector may be found in the ". . . better planting systems along with a greater intensity in manpower per hectare and to a better use of supplies." See Pollitt (1981), p. 30.

55. An analysis of the problems still affecting Cuban agricultural production yields may be found in Guzman (1984), pp. 216–218.

56. For this point see CIEM (1983), Chap. 3, sub-section 3.3.

57. See Pollitt (1981).

58. See Chonchol (1963), p. 88.

59. See Ravenet and Hernández (1984), p. 135.

60. See Comité Estatal de Trabajo y Seguridad Social (1983), pp. 136–146.

61. CEE (1981b), p. 176.

62. *Ibid.*, p. 185.

63. *Ibid.*, p. 176.

64. *Ibid.*

65. *Ibid.*, pp. 249–252.

REFERENCES

Acosta, J., "La estructura agraria y el sector agropecuario al triunfo de la Revolución," *Revista Economía y Desarrollo*, No. 9 (Cuba: January–February 1972a).

Acosta, J., "Las leyes de reforma agraria en Cuba y el sector privado campesino,"*Revista economía y Desarrollo*, No. 12 (Cuba: July–August 1972b).

Acosta, J., "La revolución agraria en Cuba y el desarrollo económico," *Revista Economía y Desarrollo*, No. 17 (Cuba: May–June 1973a).

Acosta, J., "Cuba: De la neocolonia a la construcción del socialismo (II)," *Revista Economía y Desarrollo*, No. 20 (Cuba: November–December 1973b).

"Agricultura y desarrollo económico," *Revista del BNC*, No. 5 (May 1956) cited in CEE (1981).

"Agricultura y ingreso nacional," *Revista del BNC*, No. 11 (November 1955) cited in CEE (1981).

Anuario Azucarero de Cuba (Havana: 1963).

Banco Nacional de Cuba (BNC), *Desarrollo y Perspectivas de la Economía Cuba* (Havana: 1985).

Bondarchuk, V., *Tendencias Fundamentales de la Política Agraria en el Período de Transicíon*, Vol. 3 of *América Latina: Estudios de Científicos Soviéticos: La História de Cuba* (Moscow: 1980).

Brundenius, C., *Revolutionary Cuba: The Challenge of Economic Growth with Equity* (Boulder, CO: Westview, 1984).

Castro, Fidel, *La História me Absolverá* (Havana: 1973).

Castro, Fidel, *Discurso en el XXV Aniversario de la Reforma Agraria*, 17 May 1984 (Cuba: Periódico Granma, 19 May 1984).

Castro, Fidel, *Informe Central: Tercer Congreso del Partido Comunista* (Havana: 1986).

CEDEM, *La Población de Cuba* (Havana: 1976).

Chonchol, J., *Projecto de Plan Quinquenal para el Desarrollo de la Agricultura Cubana 1961-65* (Havana: 1960).

Chonchol, J., "Analisis crítico de la reforma agraria cubana," *Revista El Trimestre Económico*, No. 117 (Mexico: January-March 1963).

CIEM, *Estudio acerca de la Erradicación de la Pobreza en Cuba* (Havana: September 1983).

Comité Ejecutivo del Consejo de Ministros, "Reglamento del Mercado libre campesino," *Revista Cuestiones de la Economía Planificado*, No. 3 (Cuba: May-June 1980).

Comité Estatal de Estadisticas (CEE), *Anuario Estadistico de Cuba* (Havana: 1975).

CEE, *Anuario Estadístico de Cuba* (Havana: 1976).

CEE, "Algunas concepciones sobre el desarrollo de Cuba en la década de 1950," Vol. 3 (Havana: 1981a).

CEE, *Censo de Población y Viviendas*, Vol. 16 (Cuba: 1981b).

CEE, *Cuba Desarrollo Económico y Social durante el período 1958-1980* (Havana: 1981c).

CEE, *Guia Estadística* (Havana: 1981d).

CEE, *Anuario Estadístico de Cuba* (Havana: 1982a).

CEE, *Cuba en Cifras* (Havana: 1982b).

CEE, *La Economía Cubana* (Havana: 1982c).

CEE, *La Economía Cubana* (Havana: 1983).

CEE, *Anuario Estadístico de Cuba* (Havana: 1984a).

CEE, *Cuba en Cifras* (Havana: 1984b).

CEE, *La Economía Cubana* (Havana: 1984c).

CEE, *La Economía Cubana* (Havana: 1985).

Comité Estatal de Trabajo y Seguridad Social, *24 Años de Revolución en la Seguridad Social Cubana* (Havana: 1983).

DOR del Comité Central del Partido Comunista de Cuba, *Los Campesinos Cubanos y la Revolución* (Havana: 1975).

Figueroa, V., and L. A. Garcia de la Torre, "Apuntes sobre la comercialización agrícola no estatal," No. 83 (Cuba: November-December 1984).

Figueras, M., "Aspectos y problemas del desarrollo económico cubano," *Revista Nuestra Indústria Económica*, No. 11 (Cuba: 1965).

Guzman, A., "Discurso de clausura del II Evento Científico de la ANEC del area de Ciencias Económicas de la UH," *Revista Economía y Desarrollo*, No. 80 (Cuba: May-June 1984).

JUCEPLAN, *Boletín Estadistico* (Havana: 1971).

JUCEPLAN, *Anuario Estadístico de Cuba* (Havana: 1972).

JUCEPLAN, *Anuario Estadístico de Cuba* (Havana: 1973).

Kalecki, M., *Hypothetical Outline of the Five Year Plan 1961-66 for the Cuban Economy* (Havana: 1960).

Lineamientos Económicos y Sociales para el Quinquenio 1981-85 (Havana: 1981).

McEwan, A., *Revolution and Economic Development in Cuba* (London: Macmillan, 1981).

Martin, A., *La ANAP: 20 Años de Trabajo* (Havana: 1982).

Martinez, F., "La Ley 3 de la Sierra Maestra y la política agraria del ejército rebelde," *Revista Economía y Desarrollo*, No. 49 (Cuba: September-October 1978).

del Monte, A., "La integración agroindustrial azucarera," *Revista Cuestiones de la Economía Planificada*, No. 8 (Cuba: March-April 1981).

Pjano Santos, O., *El Imperialismo Norteamericano en la Economía de Cuba* (Havana: 1960).

Plataforma Programática del Partido Comunista de Cuba (Havana: 1976).

Pollitt, B., "Estudios acerca del nivel de vida rural en la Cuba pre-revolucionaria: Un analisis crítico," *Revista Teoría y Práctica*, No. 42/43 (Cuba: 1967).

Pollitt, B., "La Revolución y el modo de producción en la agricultura cañera de la economía cubana, 1959-81," *Revista Cuestiones de la Economía Planificado*, No. 11 (Cuba: September-October 1981).

Primer Forum Nacional de la Ley de Reforma Agraria, Fifth Session (Havana: July 1959).

"La producción agrícola y animal en Cuba," *Informe de la Delegación de Cuba al XIV Período Sesiones de la Conferencia de la FAO*, Chap. 2 (Havana: n.d.).

Ramírez, J., "El sector cooperativo en la agricultura cubana," *Revista Cuba Socialista*, No. 11 (Cuba: June-August 1984).

Ravenet, M., and J. Hernández, *Estructura Social y Transformacciones Agrarias en Cuba* (Havana: 1984).

Rodríguez, C. R., "Cuatro años de Reforma Agraria," *Letra con filo*, Vol. 2 (Havana: 1983a).

Rodríguez, C. R., "Cuba en el tránsito al socialismo," in *Letra con filo*, Vol. 2 (Havana: 1983b).

Rodríguez, C. R., "La revolución cubana y el campesinado," *Letra con filo*, Vol. 2 (Havana: 1983c).

Rodríguez, J. L., "Política económica de la Revolución Cubana, 1959-60," *Revista Economía y Desarrollo*, No. 54 (Cuba: July-October 1979).

"Tésis sobre las directivas para el desarrollo económico y social en el quinquenio 1976-80," *Tésis y Resoluciones* (Havana: Primer Congreso de Partido Comunista de Cuba, 1976).

"Una evaluación de la reforma agraria en Cuba," *Revista Economía y Desarrollo*, No. 11 (Cuba: May-June 1972).

Yañez, E., "El SDPE y el perfecioamiento agropecuarias," *Revista Cuestiones de la Economía Planificada*, No. 7 (Cuba: January-February 1981).

Zanetti, O., *El Comercio Exterior en la República Neocolonial* (Havana: 1975).

APPENDIX

Table A1. *Distribution of land use* (in percentages)

	1946	1957	1970	1975	1984
Total surface	100.0	100.0	100.0	100.0	100.0
Cultivated	21.7	22.5	48.7	39.9	40.0
Non-cultivated pastures	42.9	44.1	22.3	21.8	15.0
Forested	13.9	12.3	6.1	21.8	29.0
Idle and other non-agricultural lands	21.5	21.1	22.9	16.5	16.0

Sources: C. R. Rodríguez (1983a), Vol. 2, p. 229; CEE (1984b), p. 62.

Table A2. *Estimated gross production value for the government agriculture and livestock sector* (in millions of pesos at constant prices)

	Sector* total	Sugar cane agriculture	Other agriculture excluding sugar cane	Cattle raising	Agriculture and livestock services†
1962‡	915.2	259.8	307.0	301.1	18.2
1963	880.1	228.9	287.5	316.8	23.5
1964	934.6	258.4	271.0	352.9	25.2
1965	984.5	318.4	222.7	334.0	29.3
1966	1,008.6	268.6	297.5	393.3	48.6
1967	1,065.8	352.9	290.6	360.6	61.7
1968	1,048.2	394.2	309.3	314.5	30.2
1969	1,225.5	406.3	274.6	298.2	247.4
1970	1,164.6	559.2	260.4	345.0	—§
1971	1,113.8	471.0	267.7	327.8	47.3
1972	1,170.0	457.5	302.3	362.9	47.3
1973	1,219.2	482.1	315.3	373.9	47.3
1974	1,268.0	480.2	364.5	376.0	47.3
1975	1,343.9	518.2	378.3	407.1	47.3
1976	1,448.6	516.7	443.0	430.5	58.4
1977	1,502.3	538.6	417.6	443.4	57.7
1978	1,591.3	634.3	434.7	463.7	58.5
1979	1,611.9	634.4	437.3	480.5	59.7
1980	1,587.6	610.6	430.0	485.6	61.4
1981	1,791.8	701.3	498.5	533.5	58.5
1982	1,692.6	654.5	489.1	490.8	58.2
1983||	1,675.1	633.1	450.3	532.7	59.0
1984||	1,733.2	643.8	486.0	540.6	62.8
1985	1,741.9	621.3	519.5	535.2	71.2
Average annual growth rates (1962–85)	2.8%	3.9%	2.3%	2.5%	6.1%

*Homogeneity of the sector's output (and consistency for the series) was attained by excluding fishing and forestry activities, which were included from 1962 to 1975.
†Agriculture and livestock services were taken based on five year averages between 1971 and 1975.
‡Data between 1962 and 1965 was estimated in accordance with the proportion of each subsector in the total sectoral production, based on JUCEPLAN data (1972 and 1973).
§1970 data for agriculture and livestock services was not available.
\|\|1983 and 1984 data was estimated from growth rates reported in CEE (1983 and 1984).
Sources: BNC (1975, p. 23); JUCEPLAN (1971, p. 46; 1972, p. 31; 1973, p. 35); CEE (1976; 1981c, p. 60; 1981d, p. 5; 1982a, p. 207; 1982b, p. 31; 1982c, p. 31; 1983, p. 13; 1984a, p. 186; 1984c, p. 11; 1985, p. 17).

Table A3. *Cane sugar production by sectors*

"Zafras"*	Production (millions of metric tons) Total	Government	Non-government	Yield by hectare (metric tons) Total	Government	Non-government
1961/1962	36.7	14.7	22.0	32.0	35.0	31.0
1962/1963	31.4	13.7	17.7	30.0	31.0	29.0
1963/1964	37.2	24.9	12.3	37.0	39.0	34.0
1964/1965	50.7	35.2	15.5	48.0	48.0	47.0
1965/1966	36.8	29.3	10.5	38.0	38.0	37.0
1966/1967	50.9	37.6	13.3	49.0	49.0	48.0
1967/1968	41.5	30.1	11.4	42.0	41.2	44.2
1968/1969	41.7	30.4	11.3	44.2	42.8	48.2
1969/1970	81.5	62.5	19.0	55.8	54.7	59.9
1970/1971	52.2	41.0	11.2	41.7	41.1	44.1
1971/1972	44.3	35.6	8.7	37.5	37.1	39.1
1972/1973	48.2	39.3	8.9	45.0	44.4	47.1
1973/1974	50.4	41.5	8.9	45.6	45.0	48.7
1974/1975	52.4	42.8	9.6	44.4	43.6	48.0
1975/1976	53.8	44.0	9.8	44.0	42.7	50.3
1976/1977	60.4	48.4	12.0	53.1	51.1	62.8
1977/1978	69.7	57.2	12.5	56.3	55.3	61.2
1978/1979	77.3	63.8	13.5	58.9	57.8	64.6
1979/1980	64.0	53.4	10.6	46.0	45.2	50.5
1980/1981	66.6	54.5	12.1	55.1	53.8	61.3
1981/1982	73.1	60.2	12.9	55.1	53.9	61.0
1982/1983	69.7	54.9	14.8	58.0	56.7	63.6
1983/1984	77.4	63.2	14.2	57.4	57.3	57.6

Sources: JUCEPLAN (1971), p. 68;
CEE (1984a), p. 192.
*Sugar cane harvesting and sugar production period.

Table A4. *Non-sugar cane agricultural production harvest by sectors*
(in thousands of metric tons)

	1962	1965	1970	1975	1980	1984
Tubers and roots	345.2	281.3	135.6	328.6	606.3	588.4
Government	95.4	126.5	75.1	183.1	391.2	393.2
% represented	27.6	45.0	55.4	55.7	64.5	66.8
Private	249.8	154.8	60.5	145.5	215.1	195.2
% represented	72.4	55.0	44.6	44.3	35.5	33.2
Cereals	311.1	71.5	307.9	358.8	453.2	542.4
Government	183.7	36.4	294.1	327.9	463.8	482.2
% represented	59.0	50.9	95.5	91.4	96.4	88.9
Private	127.4	35.1	13.8	30.9	16.4	60.2
% represented	41.0	49.1	4.5	8.6	3.6	11.1
Beans	29.1	10.8	50.0	4.8	4.1	3.9
Government	15.9	6.9	2.5	1.7	2.4	0.6
% represented	54.6	63.9	50.0	35.4	58.5	15.4
Private	13.2	3.9	2.5	3.1	1.7	3.3
% represented	45.4	36.1	50.0	64.6	41.5	84.6
Vegetables	—	273.6	156.9	447.9	419.1	502.4
Government	—	83.2	91.4	194.6	130.0	165.4
% represented	—	30.4	58.3	43.5	31.0	32.9
Private	—	190.4	65.5	253.2	289.1	337.0
% represented	—	69.6	41.7	56.5	69.0	67.1
Fruits	365.6*	393.1	330.2	578.4	770.9	978.7
Government	96.7*	126.8	192.4	341.6	530.2	753.1
% represented	26.4	32.3	58.3	59.1	68.8	76.9
Private	268.9*	266.3	137.8	236.8	240.7	225.6
% represented	73.6	67.7	41.7	40.9	31.2	23.1
Coffee	—	23.9	19.7	17.8	19.0	22.2
Government	—	4.0	5.3	8.1	11.1	13.4
% represented	—	16.7	26.9	45.5	58.4	60.4
Private	—	19.9	14.4	9.7	7.9	8.8
% represented	—	83.3	73.1	54.5	41.6	39.6
Cacao	2.4	2.2	1.3	1.2	1.3	1.9
Government	—	0.1	0.1	0.2	0.4	0.7
% represented	—	4.5	7.7	16.7	30.8	36.8
Private	2.4	2.1	1.2	1.0	0.9	1.2
% represented	100.0	95.5	92.3	83.3	69.2	63.2
Tobacco	—	35.1	31.7	41.4	8.1	44.2
Government	—	4.3	4.8	7.4	3.7	11.3
% represented	—	12.3	15.1	17.9	45.7	25.6
Private	—	30.8	26.9	34.0	4.4	32.9
% represented	—	87.7	84.9	82.1	54.3	74.4

Sources: JUCEPLAN (1971), pp. 96–104;
CEE (1984a), pp. 204–206.
*1963 data.

Table A5. *Selected livestock products* (in thousands of metric tons)

	1962	1965	1970	1975	1980	1984
Cattle	—	307.4	340.8	218.8	279.5	287.8
Government	—	248.6	—	—	278.9	287.2
% represented	—	80.9	—	—	99.8	99.8
Private	—	58.8	—	—	0.6	0.6
% represented	—	19.1	—	—	0.2	0.2
Milk	217.2*	231.3	214.4	454.3	695.4	779.1
Government	76.1*	145.9	187.1	431.8	679.9	764.4
% represented	35.0	63.1	87.3	95.0	97.8	98.1
Private	141.1*	85.4	27.3	22.5	15.5	14.7
% represented	65.0	36.9	12.7	5.0	2.2	1.9
Pork	6.2	18.4	11.9	38.4	53.8	87.6
Government	—	14.4	—	—	53.8	81.9
% represented	—	78.3	—	—	100.0	93.3
Private	—	4.0	—	—	—	5.7
% represented	—	21.7	—	—	—	6.5
Poultry	24.8	34.2	20.4	55.9	58.6	78.9
Government	20.9	30.6	20.2	55.6	58.6	78.9
% represented	84.3	89.5	99.0	99.5	100.0	100.0
Private	3.9	3.6	0.2	0.3	—	—
% represented	15.7	10.5	1.0	0.5	—	—
Eggs	174.6	920.3	1,402.7	1,748.9	2,119.7	2,344.9
Government	141.7	880.3	1,402.5	1,748.2	2,119.7	2,344.9
% represented	81.1	95.7	99.98	99.95	100.0	100.0
Private	32.9	40.0	0.2	0.7	—	—
% represented	18.9	4.3	0.02	0.05	—	—
Bee honey	2.7*	3.9	4.5	6.5	0.55	0.3
Government	—	0.4	3.0	4.6	0.08	0.2
% represented	—	10.3	66.7	70.8	14.5	67.0
Private	2.7*	3.5	1.5	1.9	0.47	0.1
% represented	100.0	89.7	33.3	29.2	29.2	33.0

Sources: JUCEPLAN (1971), p. 123;
CEE (1984a), p. 277.
*1963 data.

Table A6. *Average monthly salary paid to state agriculture workers* (in pesos)

1966	91.0*
1971	110.3*
1975	128.6*
1980	127.0
1984	173.0

Sources: Acosta (1973b), p. 95;
JUCEPLAN (1971), p. 50;
CEE (1975), p. 45;
CEE (1984a), p. 107.
*Estimated data.

Table A7. *Relative growth of agriculture production by branch and forms of ownership* (in 1965 constant prices)

	1975	1980	1984	Annual average 1975–84
				%
Agriculture and livestock sector	100	113.4	123.8	2.4
Government	100	119.7	128.3	2.8
Private	100	94.3	110.2	1.1
Sugar cane agriculture	100	118.9	125.4	2.5
Government	100	120.0	123.8	2.4
Private	100	113.6	132.6	3.2
Non-sugar cane agriculture	100	112.1	115.5	1.6
Government	100	112.3	119.9	2.0
Private	100	84.4	108.0	0.9
Cattle	100	119.2	132.8	3.2
Government	100	128.5	146.1	4.3
Private	100	93.0	95.2	−0.5
Agriculture and livestock services	100	105.5	107.9	0.9
Government	100	105.5	107.9	0.9

Source: Computations based on CEE (1984a), p. 186.

·3·

Gender Issues in Contemporary Cuban Tobacco Farming

JEAN STUBBS

1. INTRODUCTION

In the market economy of 20th-century prerevolutionary Cuba, much traditional farming had been disrupted by large-scale foreign and local capital investment. This was particularly true of sugar plantation agriculture and land-extensive cattle ranching and held to a lesser degree for tobacco and other branches of agriculture. Nowhere, however, had it ushered in any uniform agricultural modernization. On the contrary, it had often served to strengthen archaic forms of production. There were significant variations from sector to sector; operating alongside modern farming units with salaried laborers, was a particularly intensive form of small-scale labor-intensive, small-tenant and subtenant farming and sharecropping. This produced a semi-peasantry/semi-proletariat which farmed small plots of land according to an intricate system of land tenure and rent-in-kind, was highly dependent on unpaid family labor and was forced, at certain times of the year, to sell its labor.

This was particularly true of tobacco, once Cuba's second major industry and crop.[1] Nineteenth and 20th-century local and foreign capital — with some notable exceptions — preferred to use credit and buying mechanisms rather than to farm the land directly. The drive to accumulate capital was attempted through local landlords renting and subrenting their land and complex systems of sharecropping, which usually required sharecroppers to hand over one-third, one-quarter or one-fifth of the crop. A visibly exploitative form of male labor in male-headed households was accompanied by intensified women's participation in subsistence production and family reproduction crucial to family survival, plus the seasonal harvesting and sorting of tobacco.[2]

The inner workings of this essentially subcapitalist system of production and the nature of the post-1959 transition to new farming systems under the socialist-oriented agricultural development policies, are explored in this paper. The paper examines gender issues in the context of farming policies and systems, especially the new cooperative farms, in two predominantly tobacco-growing areas: San Luis (Vuelta Abajo) in Pinar del Río province and Cabaiguán (Vuelta

---

*This is a revised version of a paper presented at the Conference on Gender Issues in Farming Systems Research and Extension at the University of Florida, Gainesville, February 26–March 1, 1986. It incorporates an earlier paper "Rural tobacco women in Cuba: San Luis and Cabaiguán, 1940–80," presented at the 17th Annual Conference of Caribbean Historians in Havana, April 1985. A more general essay was written jointly with Mavis Alvarez, of the Cuban National Association of Small Farmers (ANAP) (Alvarez and Stubbs, 1986). The tobacco research was facilitated by a postdoctoral award from the Committee on Latin America Studies of the Social Science Research Council and the American Council of Learned Societies, with cooperation in Cuba from ANAP, the state enterprise CUBATABACO, the Ministry of Agriculture, the Physical Planning Institute, national and local archives, libraries and museums, but most of all the tobacco-growing families who opened up their lives and homes to me on field trips.

Arriba) in Sancti Spíritus province. It attempts to highlight significant differences between the two provinces in history, land structure and organization, and type of tobacco grown — all factors which have had a bearing on past and present gender patterns. The paper also attempts to show how overall policy has been successful in opening up new avenues for women but has at the same time thrown up new challenges which will demand future policy action if women are to consolidate their gains.

2. DEVELOPMENTS IN TOBACCO AGRICULTURE

It has been pointed out that the exceptional speed of the transition to socialist forms of postrevolutionary agrarian organization reflected in large measure an earlier process of agricultural development characterized by specialized commercial farming, in which the production of cash crops (especially sugar cane) for export was predominant. Cuba's highly differentiated agrarian class structure, with a large wage-earning proletariat, distinguished its prerevolutionary agrarian system from most, if not all, other agrarian societies which would subsequently experience transitions to socialist agriculture. Of those considered "economically active" in agriculture, forestry and fishing in the 1953 census in Cuba, less than 30% were classified as farmers and livestock breeders and some 60% were classified as agricultural wage workers. This has to be qualified in that the census procedures yielded statistically simplified rural occupational structures which made any clear-cut classification problematical.[3] Even in sugar, Cuba's most modernized crop, a significant semi-proletarianization combined wage and non-wage labor in agriculture with a diversity of non-agricultural work; this, along with intensive cheap labor and little technology, served to hold back the successful later development of state farms.

In tobacco, the fluidity between non-wage or peasant and predominantly wage forms of agricultural organization was accentuated. Tables 1 and 2 compare the 1945 size of farms, value of production and land tenure system in tobacco to sugar and cattle. Tables 3 and 4 reflect the patterns for tobacco broken down to the provincial and municipal levels. Table 5 reflects the predominance of tobacco in San Luis and Cabaiguán and Table 6 shows unpaid and seasonal labor on tobacco farms.

Several points become obvious from the comparison between San Luis and Cabaiguán. From Table 3 it can be seen that in San Luis the size of farms was clustered fairly evenly in the 1.0–4.9, 5.0–9.9 and 10.0–24.9 hectare brackets, whereas in Cabaiguán the balance tilted toward the larger farms, more in line proportionally with the provincial figures for both Pinar del Río and Las Villas.[4] Table 4 shows that in San Luis the greatest number of farms were sharecropped and were very small.

Farms in San Luis, however, were more lucrative than in Cabaiguán. Land yields and the superior quality of the export cigar wrapper tobacco grown there made for farms of double the value and income of those in Cabaiguán (as can be seen from Tables 5 and 6). Table 5 also shows how cattle and sugar were proportionately more important in Cabaiguán. Table 6 shows the large proportion of unpaid labor, especially in San Luis, where there was much greater crop intensity.

By catering specially to export and manufacturing interests from the 19th century on, both areas produced strong tobacco interests. San Luis, in particular, had respected tobacco growing families that made up an influential agrarian bourgeoisie which worked fertile land on a patriarchal system of local benefits. Major farms, including the American Tobacco subsidiary Cuban Land and Leaf and the Rodríguez family El Corojo estate operated through several farms in the area, almost all with sharecropping families and wage laborers and their own sorting sheds.[5] Owners and dealers exerted exclusive buying and selling rights over the crop and were able to turn the National Tobacco Growers' Association (founded 1942) into a powerful instrument of their own; a counter organization was the Sharecroppers' and Tenant Farmers' Association (1952), which attempted to protect members against the abuses of patron dependent relationships.[6]

The letting, subletting and sharecropping of land is crucial to any understanding of labor in the tobacco sector. Depending on the land tenure system, growers could be particularly vulnerable to landowners, creditors, buyers and the many middlemen and speculators. In Cabaiguán many landowners were absentee or managed far-removed parts of their cattle estates and let or sharecropped out the tobacco land. In San Luis, landowners characteristically oversaw part of their tobacco land and let or sharecropped out the rest. They often called on paid or unpaid tenant farmer or sharecropper labor for other services and had their own local stores which functioned on credit and pay chits. In this way, owners and managers bore little of the risk of what was a highly delicate and seasonal crop. In

Table 1. *Size of farms and value of production for tobacco, sugar and cattle, 1945*

Size of farms (hectares)	Total value of farm production $	Tobacco Farms reported	Tobacco Value $	Sugar Farms reported	Sugar Value $	Cattle Farms reported	Cattle Value $
All sizes	331,885,242	34,437	33,844,244	42,470	138,167,239	97,573	69,476,465
Up to 0.4	56,156	17	4,458	8	3,066	626	78,185
From 0.5 to 0.9	214,286	102	25,514	60	6,644	884	50,286
From 1.0 to 4.9	12,018,328	6,376	4,559,694	2,797	514,497	14,177	1,204,391
From 5.0 to 9.9	21,482,401	9,163	6,701,860	5,711	1,969,218	17,596	2,408,087
From 10.0 to 24.9	56,933,281	13,274	13,319,116	14,622	11,202,785	30,987	8,239,698
From 25.0 to 49.9	47,723,801	3,609	5,349,377	9,322	17,208,615	16,083	8,316,374
From 50.0 to 74.9	22,456,907	867	1,589,571	3,199	8,391,029	5,620	5,157,800
From 75.0 to 99.9	13,585,467	327	773,420	1,483	5,753,088	2,691	3,424,663
From 100.0 to 499.9	82,097,403	605	1,200,118	4,380	46,193,023	7,294	20,812,640
From 500.0 to 999.9	31,981,846	57	212,249	600	18,159,772	1,023	10,413,747
From 1000.0 to 4999.9	34,357,156	32	92,730	262	22,586,586	518	7,820,023
5000.0 or more	8,978,210	8	16,157	26	6,178,916	74	1,598,571

Source: Taken from Ministerio de Agricultura (1951), Table 47.

Table 2. *Land tenure of farms and value of production for tobacco, sugar and cattle, 1945*

Type of land tenure	Total value of farm production $	Tobacco Farms reported	Tobacco Value $	Sugar Farms reported	Sugar Value $	Cattle Farms reported	Cattle Value $
All kinds of tenure	331,885,242	34,437	33,844,244	42,470	138,167,239	97,573	69,476,465
Landowner	85,843,376	6,730	6,457,255	10,508	22,307,752	29,605	28,469,077
Manager	50,777,150	726	1,080,951	2,418	26,030,487	5,668	14,139,164
Tenant farmer	126,564,089	8,895	7,672,979	20,973	72,125,463	30,940	20,222,477
Subtenant farmer	13,891,384	1,547	967,396	3,266	7,399,205	4,980	1,799,347
Sharecropper	45,627,290	15,820	17,353,101	4,358	8,635,651	18,755	3,736,856
Squatter	6,520,472	553	152,781	501	186,399	6,622	756,752
Others	2,660,851	166	159,781	346	1,482,282	1,003	352,795

Source: Taken from Ministerio de Agricultura (1951), Table 48.

Table 3. *Tobacco farms according to size and territory, 1945*

	Total no. of farms	Up to 0.4	0.5–0.9	1.0–4.9	5.0–9.9	10.0–24.9	25.0–49.9	50.0–74.9	75.0–99.9	100.0–499.9	500.0–999.9	1000.0–4999.9	5000+
Cuba	159,958	1148	1877	29,170	30,335	48,778	23,901	8,517	3,853	10,433	1,442	780	114
Pinar de Río prov.	23,030	117	68	3,997	6,149	8,667	2,303	575	258	647	121	108	20
San Luis mun.	981	5	7	297	262	298	67	18	1	22	2	1	1
Las Villas prov.	40,182	361	375	6,504	7,296	12,895	6,604	2,268	1,065	2,552	313	130	19
Cabaiguán mun.	2,073	—	5	558	367	701	264	60	23	61	3	1	—

Source: Compiled using figures from Table VIII, Ministerio de Agricultura (1951).

Table 4. No. of tobacco farms and area worked under various kinds of land tenure, 1945

	Total no. of farms	Total land area (hectares)	Landowner No. farms	Landowner Total area	Manager No. farms	Manager Total area	Tenant farmer No. farms	Tenant farmer Total area	Subtenant farmer No. farms	Subtenant farmer Total area
Cuba	159,958	9,077,086	48,792	2,958,964	9,342	2,320,445	46,048	2,713,130	6,987	21,521
Pinar del Río prov.	23,030	968,853	3,373	208,114	616	200,452	4,942	225,804	1,048	21,410
San Luis mun.	981	25,109	141	3,382	24	4,491	99	10,554	31	562
Las Villas prov.	40,182	2,033,190	11,546	693,986	1,949	3,375,636	15,860	972,794	2,676	64,730
Cabaiguán mun.	2,073	43,969	504	13,110	25	4,647	656	18,509	109	1,430

	Sharecropper No. farms	Sharecropper Total area	Squatting No. farms	Squatting Total area	Others No. farms	Others Total area	Idle No. farms	Idle Total area
Cuba	33,064	552,079	13,718	244,589	2,007	72,134	636	25,210
Pinar del Río prov.	12,559	189,209	393	2,536	99	21,328	99	2,628
San Luis mun.	617	5,871	52	212	17	37	1	54
Las Villas prov.	7,166	110,692	632	7,136	349	8,218	107	10,857
Cabaiguán mun.	772	6,240	3	26	4	185	1	20

Source: Compiled using figures from Table IX, Ministerio de Agricultura (1951).

Table 5. *Value of farm production for tobacco, sugar and cattle according to territory, 1945*

	Total value of farm prod.	Tobacco Farms reported	Value $	Source of Income Sugar Farms reported	Value $	Cattle Farms reported	Value $
Cuba	331,885,242	34,437	33,844,214	42,470	138,167,239	97,573	69,476,465
Pinar del Río prov.	37,510,845	17,387	18,833,844	1,322	4,377,146	14,458	3,784,949
San Luis mun.	4,175,069	1,387	4,574,512	471	97,496	312	84,739
Las Villas prov.	77,479,694	11,833	11,031,142	16,916	32,726,832	29,379	16,392,066
Cabaiguán mun.	4,472,944	1,380	2,197,199	614	280,979	1,384	976,015

Table 6. All farms and predominantly tobacco farms according to territory, income, number of workers and wages paid, 1945

		No. of farms	Total Income	Percentage of total farm income	Unpaid All year	Unpaid Part year	Paid All year	Paid Part year	Total wages
Cuba	All farms	159,958	331,885,247	—	331,724	20,561	53,693	423,690	109,443,834
	Tobacco	22,750	40,755,323	86.6	64,395	4,994	24,965	279,155	71,503,935
Pinar del Rio	All farms	23,030	37,510,845	—	67,885	969	4,919	35,198	7,296,130
	Tobacco	12,116	21,653,249	81.4	27,572	561	1,072	15,936	2,051,235
San Luis	All farms	931	4,175,069	—	2,499	59	259	3,421	678,831
	Tobacco	816	3,936,466	94.8	2,181	57	205	3,194	636,335
Las Villas	All farms	40,182	77,479,694	—	83,999	5,914	11,500	78,789	21,362,236
	Tobacco	8,032	14,227,421	69.6	20,348	772	901	3,056	916,936
Cabaiguán	All farms	2,073	4,247,944	—	4,172	125	325	2,230	416,063
	Tobacco	1,154	2,703,314	78.8	2,582	23	202	897	201,359

Source: Taken from Ministerio de Agricultura (1951), Table 54.

San Luis, where little else was grown, tobacco families were particularly susceptible to market changes. Wage laborers, even at peak harvest times, faced an influx of migrant labor from surrounding areas, which undercut already low casual wage rates. In Cabaiguán, there was a greater chance of other agricultural and non-agricultural work, but the area as a whole was less prosperous.[7]

The fluidity of labor patterns emerged over and over again in life histories of tobacco farmers in the two areas, and this was carried over in the postrevolutionary period. If land reform and the transition to socialist agriculture in sugar had its complexities, in tobacco it had even more.

The First Agrarian Reform Law of May 1959, substantially implemented by summer 1960, set a ceiling of approximately 400 hectares on the size of private farms. Generally speaking, land from plantations above this size was taken over by the state, while land that had been parcelled out under the various farming systems was turned over to tenant and subtenant farmers and sharecroppers. These then became private farmers in their own right and were grouped together under the National Association of Small Farmers (ANAP) in May 1961. The Second Agrarian Reform Law of 1963, which limited private farms to 67 hectares, was largely motivated by a stepped up depreciation of property, capital, and services, as well as an overt hostility to what was already a clearly defined socialist process on the part of the middle agrarian bourgeoisie.[8]

Roughly speaking, the first law affected 70% of agricultural land, of which 40% became state controlled and 30% was placed in the hands of a small peasantry, leaving the remaining 30% in the hands of a middle peasantry. This last bit of private land was for all intents and purposes eliminated under the Second Agrarian Reform Law. In the late 1960s, the sale or rental of private land to the state resulted in an 80 to 20% ratio of state to private agricultural land; these figures roughly hold true today. In the late 1960s there were 400,000 laborers working in agriculture on state farms in the region and some 250,000 ANAP members; today the comparable figures are 700,000 and 180,000.

In practice, just as before the Revolution, the categories of state farm worker and small farmer have never been as neat as may have appeared. Many small farmers were, from the very early years, involved in mutual aid organizations that were given explicit support through ANAP. Peasant associations (ACs) were the most loosely organized variant, emerging originally out of prerevolutionary struggles against peasant eviction. Credit and service cooperatives (CCSs) were a more structured attempt to collectively organize agricultural inputs and services. Agricultural societies (SAs) were a further step toward actually pooling the land in the early 1960s. All helped modify traditional peasant relations of production. Conversely, state farm workers privately tended small siphoned-off plots for subsistence production. There were also constant mobilizations of casual and seasonal labor, both paid and voluntary, from state to non-state sectors and vice-versa, making for a continued, if different, fluidity of labor.[9]

The vast proportion of state land was in the sugar sector, which had been overwhelmingly a plantation economy. In other branches of agriculture, where this had been less so, the proportion of state land was markedly less. State land was at its lowest in tobacco, almost the inverse of national figures, with 25% state and 75% private land; this is explained by the prerevolutionary agrarian structure in tobacco. Table 7 shows how predominant the non-state sector has been in tobacco, up to the present day.

Again, there were variations in the two areas under study. In San Luis, a substantial part of quality leaf production fell into state hands when Cuban Land and Leaf and El Corojo lands were merged to become the new Santiago Rodríguez state tobacco farm. Also, the Patricio Lumumba experimental tobacco farm was set up on what was previously scrup land but with good soils and nearby water. In Cabaiguán, the tobacco that fell into state hands was much more dispersed, often by default rather than because it had been farmed by any large-scale enterprise before. The contrast between the San Luis and Cabaiguán experience is an interesting one and points to the question of continuity and break in land patterns and farming practices. Whereas state farms in San Luis continued the large company farming tradition, in Cabaiguán traditional growers remained in the private sector and state farms were left to grow tobacco with less experienced laborers and no such company tradition. Although various state farms grew tobacco over the years, it was never very successful, and in 1983 was left entirely to the private sector.[10] Table 8 shows the 1984 distribution between private and state sectors among tobacco farms in San Luis and Cabaiguán.

As the prime beneficiaries of land redistribution in the private sector, tobacco growers were the pioneers in peasant societies and credit and service cooperatives in the early years. They contributed substantially to the 346 agricultural societies and 587 credit and service cooperatives set up in 1963, grouping together farmers for the

Table 7. *Tobacco, land area, volume and value of production, 1970–84*

Tobacco		1970 Total	1970 State	1970 Non-state	1975 Total	1975 State	1975 Non-state	1980 Total	1980 State	1980 Non-state	1984 Total	1984 State	1984 Non-state
Land area	Total	51.9	9.4	42.5	69.3	16.5	52.8	53.4	15.2	38.2	55.1	15.6	39.5
('000s hectares)	Dark	41.6	4.9	36.7	63.6	13.9	49.7	49.5	12.8	36.7	48.2	11.0	37.2
	Mild	10.3	4.5	5.8	5.7	2.6	3.1	3.3	1.8	1.5	5.0	3.0	2.0
Crop production	Total	31.7	4.8	26.9	42.3	7.5	34.8	8.2	3.7	4.5	44.6	11.3	33.3
('000s tons)	Dark	27.6	2.9	24.7	37.9	5.8	32.0	6.7	2.9	3.9	39.4	8.4	31.0
	Mild	4.2	1.9	2.3	4.5	1.7	2.8	1.5	0.8	0.7	4.4	2.1	2.3
Value of production ('000 000s pesos)	Total	34.4	4.4	30.0	43.9	7.2	36.7	13.2	4.8	8.4	62.5	14.6	47.9

Source: Compiled with data from CEE (1984).

Table 8. *Tobacco according to farming systems, San Luis and Cabaiguán, 1984*

	Land area (*cab*)*	CPA membership Total	M	F	%F	(*cab*)	CCS membership Total	M	F	%F	(*cab*)	State farm membership Total	M	F	%
San Luis	154	681	520	161	23.6	189	1,410	1,388	22	1.6	192	1,260	956	204	16
Cabaiguán	906	1,899	1,172	727	38.3	1,385	2,775	2,097	678	24.4	—	—	—	—	

Source: Compiled from local research data.
*One *caballería* = 13.46 hectares.

collective use of curing sheds, irrigation and machinery, credit and supplies. The overall number of societies had dropped to 136 by 1967 and 41 by 1971, although the number of credit and service cooperatives in general had grown to 1,119 by 1971.[11]

The drop in the number of societies has to be seen in the context of an initial flight from sugar and tobacco, because of the market dependency they had signified in the past. In sugar, this trend was reversed in 1963, as sugar was redefined as a necessary foreign-exchange earner for future economic investment and diversification. The resulting prioritization of sugar and its state farm model affected policy and resources in other areas of production and the larger private sectors within them. The all-out effort for a record 1970 sugar harvest particularly highlighted this: the 1969–70 tobacco harvest, for example, was only 44% of the 1965–66 harvest and dramatically indicated the need for a re-evaluation of agricultural policy and small farming.

Both the 1971 and 1977 ANAP Congresses were instrumental in this respect. The 1971 Congress initiated a period of greater state attention to branches of agriculture other than sugar. The years 1971–76 were defined as a period of "tobacco recuperation," as both private and state sectors brought tobacco production back up to previous levels.[12] In these and subsequent years, much attention has been paid to crop costing and pricing and tobacco agrotechnology.

An initial price reform acted as a major impetus toward increased tobacco production. Private small farmers had previously found it more remunerative to harvest staple crops than tobacco; these were often in short supply and could be marketed both to the state and a highly lucrative black market.[13] Farm research stations (San Luis and Cabaiguán both have one in their area) and technological instititues specializing in tobacco placed an emphasis on new, improved Cuban strains of tobacco, soil improvement, irrigation, fertilizer and pesticides, new methods of curing, and technification where possible, all of which was made available to both the state and private sectors. Today, there are ongoing cooperation projects with Canada, Mexico and Bulgaria, though none grows such specialized cigar tobacco as Cuba, and this has proved a major drawback to the introduction of modern technology.

The application of technical know-how has been facilitated by generally improved educational standards in rural areas (a recent major adult education drive encouraged all state agricultural workers and ANAP members to complete a ninth-grade education). In addition to the Ministry of Agriculture's extension workers, ANAP has its own team of agronomists and local paraprofessionals, through whom there has evolved a whole technical activist movement. The extension agents encourage farmers and farm workers to come forward with improvements and innovations and to be on the phytosanitary alert. This last effect has ensured that blights such as the Blue Mold, which decimated the 1979–80 tobacco harvest, have been kept under strict control, and the volume and quality of leaf in recent years in both the state and private sectors have been excellent.

Over the years the cost of natural disaster such as hurricanes, flooding and drought were absorbed by the state, through a cancellation of debts on credits, and material and financial assistance for reconstruction. A more recent development has been the introduction of an extensive low-cost agricultural insurance scheme. And, since the 1977 ANAP Congress announced support for a pronounced pooling of private land and resources in agricultural production cooperatives (CPAs), these have received a strong boost from the state.[14]

The underlying rationale for the new CPAs, which, it was emphasized, were to be formed on a gradual, voluntary and autonomous basis, was to modernize a dispersed, labor-intensive private sector through land concentration and technification. It was hoped that productivity would increase, thereby augmenting output and undercutting first black and later free market prices on certain products.[15] At the same time, the new economic management and planning system[16] was to encourage greater decentralization of decision-making, local initiative and agricultural self-sufficiency in the state sector. In a sense, this brought about a convergence between state and private sectors, with both state farms and cooperatives in effect functioning as enterprises which rely on self-management and utilize a direct percentage of end-of-year performance for enterprise and local development.[17]

The new agricultural production cooperatives were a more carefully organized variant of attempts in the 1960s to pool private holdings to form agricultural societies (SAs). The remaining few such societies set the example by becoming CPAs; the ACs and CCSs out of which other CPAs grew were already versed in acting collectively on behalf of individual farmers: the land might still have been worked individually, but ACs and CCSs negotiated agreements on state quotas for production, inputs and credits.

In the CPAs, land and other basic means of production are collectively owned and each individual farmer's land contribution is valued and paid off over a period of time from funds set aside by the cooperatives expressly for this purpose. The cooperative is farmed and run collectively as an autonomous enterprise within the constraints of national and regional planning. It receives low interest credits from the state and preferential treatment in the allocation of certain resources. A percentage of profits goes to the state in return for services, a percentage is turned back for production and amenities, and the rest is divided among members, according to their labor contibution. The cooperative elects its own president and executive committee and meets in full once a month. At a major end-of-year meeting, production plans, investment programs, and consumption requirements are decided upon, as well as such issues as advance pay, profit sharing and the admittance of new members.

Individual farmers not wishing to join a cooperative formed in a given region are not pressured to do so. Conversely, agricultural laborers — and in exceptional cases even industrial workers — may join the cooperatives. In such cases, there is no material contribution other than labor and hence no compensation for means of production, although these members do have statutory rights identical to those of other cooperative farmers. The same applies to landless wives and grown children of male household heads.

It was envisioned that once the first few cooperatives were organized, their greater social and economic advantages would be widely recognized and others would soon be formed. The cooperative movement did, in fact, mushroom beyond all expectations[18] and agricultural output doubled and in many cases tripled. The number of cooperatives and membership in cooperatives peaked in 1983, and a less marked growth is expected in coming years. The number of CPAs has dropped since 1983 because of fusions, hence the average CPA land size has increased. Membership, which rose to over 82,000 in 1983, has fallen back to 71,000; in large part this reflects the age structure of cooperative membership and the new social security laws, whereby cooperative farmers may, for the first time, take paid vacation and retire on a pension.[19] The drop is also due to members leaving the cooperatives, a point touched on in the last section of this article.

Today tobacco CPAs account for some 50% of privately held land; they have in many cases increased their yields but are also running into problems. The cooperative movement as a whole faces problems ranging from organization, state-cooperative relations, "illicit" extra-agricultural activities and the negative impact of lucrative farmer and middleman speculation, all of which were critically analyzed at the May 1986 Second National Meeting of CPAs. In tobacco, there is also the problem of continued sharecropping in the individual farm sector, although on a much less exploitative scale, and the very key question of economies of scale.

CPAs stood to gain the most from land concentration through technification and resulting economies of scale. Given the kind of tobacco grown in Cuba, however, the introduction of technology on cooperatives, as on state farms, is generally limited. Moreover, following the general cooperative trend, the problem of increasing land size has been compounded by the problem of falling membership. Table 9 shows that, in the case of tobacco, the number of cooperatives reached a peak in 1981 and then fell back to the 1979 level by 1985, though the land area had more than quadrupled. Membership was at a peak in 1983 but had fallen such that in 1985 for four times the land area, there are only twice as many farmers.

Interestingly, the structural and financial constraints facing the cooperatives are very similar to those facing the state farms which, even when highly successful in terms of the volume and quality of the tobacco grown, are running at a considerable loss. Observation shows that the more tobacco intensive the farm, the greater the financial loss. Of a total of 138 tobacco CPAs whose 1985 financial standing has been analyzed,

Table 9. *Tobacco agricultural production cooperatives (CPA)*, 1979–84

		1979	1981	1983	1984
No. CPAs	Unit	220	233	230	220
Land area	'000s hect	28.8	55.6	116.5	128.4
Membership	Unit	6,315	9,384	15,347	12,836
Av. CPA area	'000s hect	130.8	238.9	506.6	507.0
Av. CPA members	Unit	29	40	67	58

Source: CEE (1984).

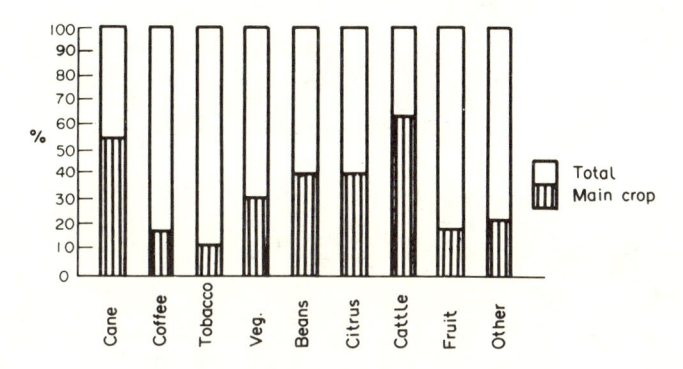

CPA land area according to main crop, 1984

51 — the majority of which were in Pinar del Río province, where the cooperatives relied much less on mixed farming and were much more tobacco-intensive — reported net losses. One core problem for both the cooperative and the state farm that helps explain these losses is the relative shortage of and high cost of labor.

One San Luis state farm manager cited as major outlays: (1) labor costs, (2) fertilizers and pesticides and (3) cheesecloth (the farm grew shade tobacco), in that order. The labor costs included bringing in temporary labor from other areas, transport, accommodation, and food costs, and higher salaries for non-agricultural workers. Cooperative farm presidents expressed similar concern over labor costs, especially finding and bringing in outside labor. The great shortage of agricultural labor is rooted in the rapid development of other sectors the economy which compete too favorably with agriculture. It is also, however, very much related to the transition from household to collective (whether state or cooperative) farming in that household farming has traditionally had a high family labor component.

3. WOMEN'S LABOR IN TOBACCO

The otherwise very complete 1946 Agricultural Census has virtually no separate figures on women and is symptomatic of Cuban statistics in general as far as information on women's work is concerned. The census recorded over 800,000 people as working in agriculture or 41.5% of the economically active population of 14 years and over. Supposedly included in this were farmers, agricultural laborers and unremunerated family labor, and yet women constituted only 1.5% of the estimated economically active population working in agriculture. According to the 1953 Population Census, less than 2% of those working in agriculture were women, the vast majority of whom worked as wage laborers rather than unwaged family laborers.

Figures such as these have been used to show the low involvement of Cuban women in prerevolutionary agriculture; they have been explained in terms of the prerevolutionary market economy and the disruption of traditional farming, which generated a surplus of rural labor and kept subsistence farming and the work of

Table 10. *Agricultural production cooperatives (CPA) in tobacco, sugar and cattle, 1984*

Region		No. CPAs	Total no. of members	Total land area (*cab*)*	Av. CPA land area (*cab*)*	Land area per member (*cab*)*	Area given to main activity (%)	Av. CPA membership
Tobacco	Cuba	220	12,836	8,311.12	37.78	0.65	12.85	58
	Pinar del Rio prov.	139	7,404	4,670.47	33.60	0.63	15.21	53
	Sancti Spiritus prov.	29	2,602	1,572.81	54.23	0.60	11.30	90
Sugar	Cuba	433	31,449	29,431.81	67.97	0.94	52.54	73
	Pinar del Rio prov.	12	683	542.85	45.24	0.79	54.05	57
	Sancti Spiritus prov.	26	2,031	2,013.79	77.45	0.99	44.42	78
Cattle	Cuba	192	6,020	9,944.12	51.79	1.35	66.68	31
	Pinar del Rio prov.	10	330	898.50	89.85	2.72	22.54	33
	Sancti Spiritus prov.	3	121	206.31	68.77	1.71	91.29	40

Source: Compiled with data from CEE (1984).
*One *caballería* = 13.46 hectares

women to a minimum. Just as it is more accurate to look at a semi-peasantry/semi-proletariat, so it is important to probe women's intensified participation in subsistence production and family survival, as well as seasonal harvesting and sorting. In tobacco, for example, while there was no gender breakdown by crop in the 1946 Agricultural Census, it can be deduced that a sizable part of the substantial returns on permanent unpaid and temporary paid labor in Table 6 came from women, and that a great many more women were not included in census returns.

While mitigated, it is clear that women's work not only continued after the Revolution but continued not to be recognized, least of all in official statistics. Hence, over the intercensal period of 1970–81, unpaid family labor supposedly dropped from 3.9% to 0.2% of the total economically active population and from 1.2% to 0.0% for women. By the 1980s, less than 10% of the economically active population working in agriculture were women.

In the small farm sector, women's participation continued to go unrecorded and hence largely unheeded, though women did respond in large numbers to the volunteer agricultural brigades organized jointly by ANAP and the Federation of Cuban Women (FMC).[20] In the state farm sector, factors such as increasing technification and mechanization changed the requirements demanded of labor: new skills, that for social reasons were not always easy for women to acquire, were needed. An initial separation of agricultural work from the domestic unit posed the classical break between "work" and the home; improved rural living standards as a result of overall development policy often meant less economic pressure on women to supplement family income and when women did choose to work, openings in such expanding non-agricultural spheres as health and education often proved far more attractive. As a result, women were to be found most in the (unremunerated) casual and seasonal work force, the (often unremunerated) volunteer brigades, and as an integral part of (unremunerated) small farmer family labor.

A breakdown of state and private sectors in the 1984 Statistical Yearbook, shows that, of 70,000 women active in agriculture in 1984, slightly over 60,000 worked in the state sector, some 5,000 were members of cooperatives, 607 worked as hired labor in the private sector, 505 were self-employed and 220 performed unpaid family labor. ANAP returns for that same year seemed to contradict the private sector figures, showing a total of some 11,000 cooperative farmwomen and just over 10,000 women members of CCSs and ACs.

In "straight" economic terms, ANAP figures are largely inflated. Membership in a cooperative did not automatically imply that a woman performed agricultural field work, and membership criteria varied from cooperative to cooperative. Some cooperatives automatically included women who contributed land (either in their own right or jointly with their husbands) whether they worked in the fields or not and were hesitant to admit women who did work but contributed no land. CCS and AC data are even more tricky to handle, since women members of small farm families expressed little need to register as members when they could be represented by their men. Those women who did register had usually become active in the local organization and this was more of a determinant than their labor.

Given the inadequacy of the general data available and the almost total absence of any gender breakdown by crop (beyond cane and non-cane agriculture), plus the fact that the conceptualization of women's economic activity was highly colored by traditional definitions, it was essential to build up some good local disaggregated data, incorporating life history techniques. In the case of tobacco, from the research conducted in San Luis and Cabaiguán,[21] it became clear that women had, indeed, been central to what was mainly small-scale tobacco production.

Historically, the domestic division of labor had been fairly complete; men oversaw agricultural production in the broader sense and rarely took part in servicing the family and household, whether it be washing, cooking, cleaning, caring for the children, fetching and carrying water, picking tubers, grinding corn, or feeding the chickens and pigs.

If this was not considered work by the census enumerators, it certainly was by peasant families, both men and women alike. The wider societal view of women in the home was very much tempered by the recognition that women's work was crucial, although within socially defined categories of what that work was. The more women's unpaid labor could be exploited, the greater the surplus that could be extracted from sharecropping families and male wage labor. The same applied to child labor, which helps explain why large familes were prevalent.

The general pattern was that girls and boys started working at age seven, the boys in the fields with their fathers and the girls around the house with their mothers. At the height of the harvest, when extra labor was essential, girls' and women's work also involved cooking for the field hands. The poorer the peasant family, the greater the need to fall back on family labor, in which case women and girls would also be out in the fields planting, weeding, pruning and harvesting the tobacco.

The type of tobacco determined to a large extent the kind and amount of work the women did. The cigar wrapper tobacco of San Luis, for example, has traditionally been harvested by leaf in baskets and the leaves then threaded together to be strung on poles to dry. This was seen almost exclusively as women's work; the drying, in particular, was done in the shade, with needle and thread. The dark filler tobacco of Cabaiguán was stickier (from the black resin) and tougher. It was traditionally harvested by knife in stalks of four leaves at a time, hung over the outstretched arm, and then transferred to poles. Heavy work such as this, and field labor in general, were traditionally considered unsuitable for women. In any case, the Cabaiguán tobacco did not need to be threaded, as it was hung straight on poles in the barns.

There were also variations in the seasonal sorting of tobacco in the initial months after harvesting, during which both municipalities provided temporary employment for thousands of women. The better quality wrapper tobacco of San Luis demanded greater classification into grades and therefore more skilled and better paid personnel; on average 100–300 people were employed in each sorting shed in peak periods. The more select the farm, the more select this, one of the few forms of paid labor for rural women, But there were many sorting sheds where the economic necessity of tobacco families was such that women and girls from age 10 would accept pittance rates for long hours of work. In Cabaiguán, conditions and pay in the sorting sheds were on the whole worse and, given the lesser intensity of tobacco growing and the concentration of larger sorting sheds (holding up to 1,000) in towns, extensive distances had to be travelled to reach the work.

With the agrarian reform and rural development policies, the land peasant families worked was made their own, the family wage was effectively upped, children were sent to school, and, despite explicit government policies to the contrary, the old societal definition of women servicing the home and family was to a certain extent reinforced, especially with the fall in tobacco production in the 1960s and the decreased labor needed in the fields and the sheds. The "tobacco recuperation" of the 1970s, including improved wages and working conditions for both men and women the FMC ANAP brigades, and regularized payroll work for women in harvesting and sorting through CUBA-TABACO, the umbrella state tobacco handling enterprise,[22] and the production cooperatives, have helped to redress this disadvantageous position of women in the labor market.

The cooperatives marked a definite policy departure as far as women were concerned. Under the initial agrarian reform of 1959, female heads of household had been accorded land titles, but they were the exception rather than the rule and often the title had meaning only on paper and not in practice, as a man would take on farm production responsibilities. In the long run, the Revolution's wider educational and agricultural development policy opened up new horizons for farm women.[23] It was not until the new CPAs, however, that a specific agricultural policy prescription for women was spelled out and women were encouraged to join cooperatives in their own right and have statutory rights identical to those of any other cooperative member.

The result, in less than a decade, has been quite startling. Following the initial flurry of cooperative formation, by 1979, over a third of all CPA members were women. In 1983, a peak year for the cooperative movement, the figure had dropped to 27%, although in absolute terms the number of women had more than quadrupled. Today, both absolute and relative figures have dropped, again partly because of the age factor, partly because recent figures apply more strictly to working women members, that is, active rather than passive ones. According to 1985 figures, women accounted for 25% of total CPA members and 12% of executive committee members. While moderate in scale, this sudden visibility of women has outstripped the state agricultural sector, in which, in 1985, women accounted for only 14.4% of the work force and 6% of executive posts.

Tobacco areas have been among those to show higher percentages of women coop farmers. The CPAs have brought with them a marked increase in the visibility of women as compared to the rest of the private sector. In the San Luis and Cabaiguán areas, all tobacco farmers are now in CPAs or CCSs; in San Luis women comprise 24% of the membership in CPAs and 1.6% in the CCSs; in Cabaiguán the figures are 38% and 25% respectively. For the state farms in San Luis, women make up 16% of the workforce, still lower than the CPAs. In all three, the figures exclude the considerable seasonal labor of women, which at the height of the harvest can run into hundreds. For this, local non-working women and women from other sectors of production work on a contract basis through CUBA-TABACO.

Despite overall policy prescriptions, mem-

Table 11. *Private sector composition. Agricultural production cooperatives (CPA), San Luis and Cabaiguán, 1984*

CPA	Land area (cab)*	Membership Total	Membership Men	Membership Women	Membership Women % total	Land contributors Total	Land contributors Men	Land contributors Women	Land contributors Women % total	Labor force Total	Labor force Men	Labor force Women	Labor force Women % total	FMC members
San Luis														
Menelao Mora	7.4	42	30	12	28.6	13	13	—	—	ng	ng	ng	ng	ng
José Martí	11.5	52	32	20	38.5	26	26	—	—	ng	ng	ng	ng	ng
José A. Echeverría	19.7	73	51	22	30.1	28	28	—	—	ng	ng	ng	ng	ng
Frank País	8.7	42	33	9	21.4	28	28	—	—	ng	ng	ng	ng	ng
Isidro de Armas	6.1	49	43	6	12.2	31	31	—	—	ng	ng	ng	ng	ng
Niceto Pérez	10.1	52	49	3	6.1	27	27	—	—	ng	ng	ng	ng	ng
Lenin	2.9	23	22	1	4.5	11	11	—	—	ng	ng	ng	ng	ng
Leopoldo Trocha	4.4	28	23	5	2.2	15	15	—	—	ng	ng	ng	ng	ng
Alfonso Valdez	9.0	30	27	3	10.0	34	24	—	—	ng	ng	ng	ng	ng
Hnos Venas	16.2	83	54	29	35.0	18	18	—	—	ng	ng	ng	ng	ng
17 de Mayo	10.2	28	22	6	21.4	15	15	—	—	ng	ng	ng	ng	ng
Carlos Lóriga	20.5	66	49	17	25.8	27	27	—	—	ng	ng	ng	ng	ng
Antonio Guiteras	11.0	52	45	7	13.5	17	17	—	—	ng	ng	ng	ng	ng
General Antonio	13.0	49	34	15	30.6	16	16	—	—	ng	ng	ng	ng	ng
Renato Guitart	2.8	22	16	6	37.5	11	11	—	—	ng	ng	ng	ng	ng
Cabaiguán														
La Nueva Cuba	51.6	191	112	79	41.4	58	33	25	43.1	114	74	40	35.1	84
Juan González	135.8	324	176	148	45.7	98	44	54	55.1	156	128	28	17.9	148
10 de Octubre	118.2	289	177	112	38.8	104	62	42	40.4	193	125	68	35.2	110
13 de Marzo	58.1	122	82	40	32.8	37	15	22	59.5	68	52	16	23.5	40
Aramis Pérez	62.8	170	100	70	41.2	66	50	16	24.2	117	87	30	25.6	65
21 Aniversario	34.8	65	42	23	35.4	24	13	11	45.8	53	40	13	24.5	23
6ta Cumbre	30.5	66	40	26	39.4	27	19	8	29.6	31	27	4	12.9	26
Victoria de Angola	29.9	82	51	31	37.8	32	16	16	50.0	56	45	11	19.6	31
Victoria de Girón	85.9	50	32	18	36.0	30	30	—	—	43	31	12	27.9	18
La Victoria	22.4	28	23	5	17.9	18	10	8	44.4	24	20	4	16.6	—
Noel Sáncho Valladares	50.0	60	43	17	28.3	23	16	7	30.4	34	27	7	20.6	17
Mártires de Cabaiguán	32.3	73	53	20	27.4	30	21	9	30.0	43	36	7	16.3	40
Romanico Cordero	28.1	78	53	25	32.1	36	24	12	33.3	72	61	11	15.3	30
Eduardo G. Lavandero	41.5	111	66	45	40.5	48	18	30	62.5	65	52	13	20.0	45
Camilo Cienfuegos	58.8	54	33	21	38.8	36	20	16	44.4	45	35	10	22.2	8
1ro de Enero	45.4	75	49	26	34.7	40	14	26	65.0	52	42	10	19.2	80
Mártires de Neiva	20.0	61	40	21	34.4	17	3	14	82.4	44	36	8	18.2	21

Source: Compiled with data from local ANAP returns for 1984.
*One *caballería* = 13.46 hectares

Table 12. *Private sector composition, credit and service cooperatives (CCS), San Luis and Cabaiguán, 1984*

CCS	Land area (cab)*	Membership Total	Membership Men	Membership Women	Women % total	Titleholders Total	Titleholders Men	Titleholders Women	Women % total	Landless members Total	Landless members Men	Landless members Women	Women % total
San Luis													
Luis Ferrer	16.0	91	91	—	—	54	54	—	—	37	37	—	—
Jesús Menéndez	11.0	137	132	5	3.6	103	102	1	0.9	34	30	4	11.8
Calixto Sánchez	10.0	39	39	—	—	21	21	—	—	18	18	—	—
Cuco Barceló	18.0	126	115	11	8.7	101	100	1	0.1	25	15	10	40.0
Viet Nam Heróico	18.0	75	75	—	—	45	45	—	—	30	30	—	—
Sergio González	5.0	33	33	—	—	20	20	—	—	13	13	—	—
Mariana Grajales	7.0	75	75	—	—	48	48	—	—	27	27	—	—
José Maceo	13.0	122	120	2	1.6	55	55	—	—	67	65	2	3.0
Cmdte Acanda	9.0	85	85	—	—	45	45	—	—	40	40	—	—
26 de Julio	9.0	71	71	—	—	40	40	—	—	31	31	—	—
Ignacio Agramonte	7.0	48	48	—	—	29	29	—	—	19	19	—	—
José A. Labrador	19.0	122	122	—	—	61	61	—	—	61	61	—	—
Pedro Ortiz	17.0	132	130	2	1.5	68	68	—	—	64	62	2	3.1
Ramón L. Peña	4.5	32	32	—	—	25	25	—	—	7	7	—	—
Francisco Barrios	5.0	62	62	—	—	50	50	—	—	12	12	—	—
Camilo Cienfuegos	20.0	160	158	2	1.3	108	108	—	—	52	50	2	4.0
Cabaiguán													
Alfredo López Brito	28.7	81	54	27	33.3	38	36	2	5.3	43	18	25	58.1
Emilio R. Capestany	51.4	110	84	26	23.6	60	50	10	16.6	40	34	16	40.0
Nieves Morejón	62.5	130	84	46	35.4	63	49	14	22.2	67	35	32	47.8
Niceto Pérez	53.7	118	101	17	14.4	73	68	5	6.8	45	33	12	26.7
Arturo Cabrera	53.6	101	70	31	30.7	55	50	5	9.1	46	20	26	56.5
Ciro Redondo	52.6	61	50	11	18.0	37	33	4	10.8	24	17	7	29.2
Francisco Rivas	40.0	63	47	16	25.4	35	29	6	17.1	28	18	10	35.7
Patria o Muerte	46.2	96	59	37	38.5	36	32	4	11.1	60	27	33	52.4
Jerónimo Ramírez	73.7	94	81	13	13.8	67	62	5	7.5	27	19	8	29.6
Julio Piñero Lorenzo	33.0	82	49	33	35.1	32	30	2	6.3	50	19	31	62.0
Rogelio Rojas	60.9	118	77	41	34.7	55	47	8	14.5	63	30	33	52.4
Horacio González	20.0	57	34	23	40.4	22	20	2	9.1	35	14	21	60.0
Sergio Soto	72.8	148	104	44	29.7	62	56	6	9.7	86	48	38	44.2
Gerardo Abreu Fontán	25.2	64	43	21	32.8	23	22	1	4.3	41	21	20	48.9
Alfredo Ferrer López	59.4	111	78	33	29.7	61	55	6	9.8	50	23	27	54.0
Beremundo Paz	50.6	96	69	27	28.1	48	45	3	6.3	48	24	24	50.0
26 de Julio	40.0	55	36	19	34.5	25	23	2	8.0	30	13	17	56.7
Hnos Calera	50.6	90	66	24	26.6	56	51	5	8.9	34	15	19	55.9
Jorge Austiny	33.1	70	45	25	35.7	39	32	7	17.9	31	13	18	58.1
Eloises Pérez	37.8	88	70	18	20.5	66	59	7	10.6	22	11	11	50.0
Marino Rodriguez	17.4	38	32	6	15.8	25	22	3	12.0	13	10	3	23.1
Jesús Menéndez	16.5	32	27	5	15.6	24	19	5	20.8	8	8	—	—
Armando González	34.3	97	87	10	10.3	55	48	7	12.7	42	39	3	7.1
Julio Careaga	38.0	76	71	5	6.6	42	40	2	4.8	34	31	3	8.8
Frank País	54.8	103	92	11	10.7	58	56	2	3.4	45	36	9	20.0
Abel Santamaría	49.8	76	66	10	13.2	53	49	4	7.5	23	17	6	26.1
Mártires de Taguasco	95.6	140	120	20	14.3	104	94	10	9.6	36	26	10	27.8
Luis Turcios Lima	90.0	140	122	18	12.9	99	84	15	15.2	41	38	3	7.3
José Martí	51.3	79	70	9	11.4	29	22	7	24.1	50	48	2	4.0
Fernando Conde	41.0	95	68	27	28.4	58	47	11	19.0	37	21	16	43.2
Ramón Balboa	40.0	66	41	25	37.9	34	27	7	20.6	32	14	18	56.3

Source: Compiled with data from local ANAP returns for 1984.
*One *caballería* = 13.46 hectares

bership criteria differed enormously from CPA to CPA in tobacco. At one of the older cooperatives in Cabaiguán, it was found that both men and women were automatically made members upon pooling their land and membership numbers were fairly equal for the sexes. In San Luis this was not the case, as no women were quoted as land contributors. Among the women who were not considered land contributors, strict rules regarding stability in agriculture often worked against them. In all cooperatives, there were both active and non-active members of both sexes. Among the men, the non-active were usually retired. Among the women, there were found anomalous case of "landed" women members who neither worked in the fields nor were active cooperative business and women who worked substantially in agricultural production but were not landed and whose membership had not been recognized.

The lower percentage of women in cooperative members in San Luis can largely be explained by the urban setting which provides other job opportunities, in tobacco and in general. One recently formed cooperative found that there were simply no women who could join, except a young graduate accountant. Wives of cooperative farmers were already on the CUBATABACO payroll as paid workers and would be hired back out to the cooperative during the harvest time. Other wives were older or had young children and could only work during peak harvest periods. Grown daughters were often working for the state sector in education, health and the like, at least until marriage and small children interrupted work. Even then, San Luis differed from Cabaiguán in that it offered more accessible daycare facilities.

Women's work on the cooperative varied considerably. In the dark filler tobacco area of Cabaiguán, where women had traditionally been less involved in tobacco, women might be organized into support brigades in non-tobacco activities: producing root and other vegetables for local consumption and sale to the state — although this was by no means obligatory and some women took a pride in working the tobacco. In San Luis, other crops were produced less, women had traditionally worked more in tobacco, and this division of field labor was not marked. Interestingly, when Cabaiguán cooperatives last year experimented with Burley tobacco, the women were particularly pleased to be able to pick and thread the leaves. One cooperative even took the tobacco to the women in their homes to be threaded on their front porches. Women who had retired from the fields or had other family commitments were all able to help out and earn some money.

So far, household servicing is still only dimly perceived as part of the collective process of accumulation of wealth, except in a semi-conscious sense over issues such as meals. Few cooperatives organized collective lunch for their members, only for outside seasonal workers. Cooking lunch for outside workers was recognized as paid labor, cooking for family members was not. One male accountant ventured to say that collective lunch facilities were "costly" to the cooperative, while at the same time, he lamented that women's responsibilities in the home worked against their stability in field labor. At that cooperative, almost half of all members were women. Of that 50% only 19% acutally worked in the fields, and of that 19%, only 38% (or 7% of the total reported female membership) worked at all regularly. Correspondingly, women worked only 11% of the total number of days worked in the year and took a corresponding 11% of annual profits.

This was an extreme case. A nearby cooperative reported only 15% of members were women, but almost all worked on a regular basis and took a more proportionate share of annual profits. Nonetheless, the general comment was that women did not like the field work, which was clearly hard agricultural labor under a hot sun. In the Cabaiguán area, women took on sewing from a nearby garment factory as a softer option.

Younger women are increasingly coming back to work on the cooperatives, in a technical capacity — as the accountant or agronomist — a trend which is noticeable generally in the tobacco areas, where almost a third of such technical personnel are women.[24]

4. GENDER ISSUES IN FOCUS

It was a particular combination of social and economic factors of the cooperative movement which proved most attractive to Cuban women, engendering their support for cooperativization. While male peasant farmers spoke of cooperativization as a wrench to stop individual farming of the land, women, especially those in the remote areas, viewed the formation of production cooperatives and cooperative villages more in terms of access to running water, electricity and amenities such as stores and schools.[25] In this sense, women were perhaps motivated more for reasons of reproduction than production, but in the process they became part of a socio-economic unit in which individual well-being depended directly on collective economic success. Women's participation was recognized as both integral and important. The extent to which women were incorporated into production, however, varied considerably and was inevitably colored by traditional farming practices, development planning strategies, and subjective as well as objective factors at the local cooperative level.

In the tobacco areas studied, there had been a marked increase in the visibility of women and also a generally heightened awareness of women's role in production and women's subordination, even when women did not necessarily participate in either production or the day-to-day management of the cooperative. On the whole, neither the men nor the women expressed the view that women should not work outside the home for the cooperative, but spoke rather of how women's domestic responsibilities are an obstacle to their increased participation.

This is reflected in the high number of women laborers at the casual and seasonal level and the relatively low number of women at the executive level. Over the years, the FMC and the government have made a concerted political effort to probe the problems women face in taking on regular employment and securing job promotions. Over the last decade, this effort has ranged from political education work aimed at passing the new 1975 Family Code, which challenged an established gender division of labor, to campaigns against on-the-job and promotion discrimination against women.

The complexities and dimensions of women's subordination have, however, proven highly resistant to change, especially in the less developed rural areas. In the CPAs, for example, an attempt has been made to ensure that at least one woman member sits on the executive committee. Even so, women members still tend to be charged with the accounts or educational and recreational work, while the men handle production.

In the Cabaiguán areas men complained about women members because not only did they work in the fields less, they also took less of an interest in day-to-day cooperative business. For example, they claimed, some would not show up for meetings. The women had their own complaint: the men would have to change and begin to carry their weight around the home in order for women to be able to participate more.

Generational change, especially among the women, is very marked and much talked about on the CPAs. While much has still not changed, there is a growing questioning of gender roles. "Machismo" is a current topic of conversation.

The older generations feel that things are a lot easier today than in their time and that new doors are opening for women but that it is too late for them to change. The younger generations, schooled away from home, are beginning to challenge certain taboos.

Nena, for instance, now retired and in her mid-60s, remembers how from her early teens she had to help her mother carry food to the field hands and fetch and carry water. Later, she worked in the sorting sheds while also taking in washing and ironing. Now, she virtually runs her own household, as well as her daughter's next door.

Her daughter, who works in the new local store, grew up in the early years of the Revolution. She, nonetheless, was quite restricted as a girl and married at 17. The granddaughters now have opportunities denied their mother. Nena's son, who did study and become a college professor, continued to live with his mother in the cooperative village and was one of the men who did housework, even after marrying. In contrast, her daughter's husband, as president of the cooperative, has little or no time for the home. The granddaughters expressed a certain resentment at being expected to help in the home much more than their brother. The oldest granddaughter in particular was critical of old-fashioned small community gossip which limits girls' mobility.

While few men currently participate in housework, it is at least on the agenda. The women who make the greatest demands on men tend to be those who hold down jobs or have political and social responsibilities. They might also be involved in the recently formed women's militia, making theirs a double or triple shift. These women are gaining confidence and questioning established mores, including the gender division of labor, male authority and dual sexual standards.

Women expressed resentment that men still thought it their right to have affairs while expecting that women should not. When the men in question held posts of authority, this could be cause for collective action. In the case of one married couple, both on the executive committee, in which the husband went with a younger woman member, the situation was such an embarrassment for the wife and other committee women, that the man was suspended from his duties. Clearly, male-female tensions of this kind, both at a cooperative and personal level, will not be resolved in the near future. They can, however, be handled supportively when consciousness has been raised and crucial space created for women.

Special technical and advisory services from the state to the tobacco sector, both state and private, plus the recent land concentration in the form of cooperatives, with a centralization of resources and production, have reaped bumper harvests in recent years. Similarly, tobacco areas have reaped the benefits of a state investment program in social as much as economic spheres, with a considerable injection of self-help.

However, several policy imperatives related to tobacco production are now emerging. One is greater land concentration, so as to avoid further dispersal of land. In the more tobacco-intensive areas of San Luis, where there is a veritable mosaic of state, cooperative and individual private farms, all interspersed, an exchange of plots is essential for any further advance. A second policy imperative is a greater attempt to select maximum productivity work brigades and organize smaller, more manageable plots on a piece-rate basis.[26] A third is a revision of crop pricing.

All three are felt to be crucial if tobacco growing is to be made more profitable again. In the case of the cooperatives, other factors are important: the inevitable outlays required of any new investment, and the inefficiencies in organizing and managing a new kind of production unit that requires technification and new skills and is susceptible to over-ambitious production targets. This is especially likely as the cooperative grows in land size and the number of cooperative farmers drops, as has been the trend over the last few years. And yet, it was state farm managers, whose farms had been going much longer, who most strongly argued for a price review if profitability were to be achieved, especially when the exporting and manufacturing of tobacco were particularly profitable.

Perhaps a fourth policy imperative should be added: a reconsideration of the labor question in gender terms. This would imply recognizing the effect on labor costs and end-of-year profits in a crop like tobacco, a traditionally small-farm product with a high unpaid family labor component, of the transition from a household to cooperative or state farm unit, in which all crop labor is economically computed and remunerated under a socialist rather than a capitalist system, whereby there are applied minimum guarantees and wages for workers.

Interestingly, bank statistics[27] show a much lower positive end-of-year balance on credits for the cooperative tobacco farmer than for the individual tobacco farmer (140 pesos against 1510 on average, a ratio of less than 1:10, for the 1983–84 harvest). The only approximate indicator available that can serve as a point of comparison between the cooperative and individual farm, this bank statistic is in itself highly problematical, precisely because it conceals the real distribution of income in each type of economy. In the individual small farm, all income is registered under the farm owner or representative; the work of family members and others in the production process is not taken into account. In the case of the cooperative, the farmer's income is that which is received for work done individually, as well as advance daily pay, social security, paid maternity leave for women, sickness, and other similar benefits (and there may be two or more cooperative farmers in a single family).

It is to be noted that, in the case of sugar, which is much less a small farm product, has a much lower female labor input, and admits mechanization and economies of scale, the figures for cooperative and individual farmers are much closer (734 pesos against 886, a ratio of some 9:10, for 1983–84). In coffee, which shows much greater similarity to tobacco, the figures are again very uneven (40 pesos against 532, less than 1:12, for the same harvest).

To pursue the point further, it is important to note that today Cabaiguán cooperatives are more profitable than San Luis cooperatives. Table 13 shows average land size, membership and profit margins for tobacco areas. In Sancti Spíritus, coops are less dependent of tobacco and the profitability edge comes from the more extensively farmed and more mechanized cattle and sugar estates, whereas Pinar del Río relies much more exclusively on highly labor-intensive and specialized tobacco.

In the case of the cooperatives, which are conceived not only as an economic but also as a social form of organization, such factors can be crucial. They are seen as another step forward in the humanization of work in rural Cuba, and as coming from a more rational and equitable social system, as much as from straight technical advance. Cooperatives have to prove themselves from both a productive and social point of view and, while some of the older (male) farmers might be a little disheartened with the process and even desist (again Pinar del Río is proving to have the highest desertion rate — 6% of membership in 1985, as against only 1% in Sancti Spíritus and a national average of 3%), women and younger generations in particular want to hold onto the wider benefits and not turn back the clock. The crucial problem of retention in agriculture so necessary to the country depends now on the level of attraction to rural areas and that in turn involves a challenge to many traditional areas of life.

With strong policies and support for women at

Table 13. *Tobacco agricultural production cooperatives (CPA), 1984-85 partial balance ('000s pesos)*

Region	No. CPAs*	No. of members	Value of production	Cost per peso of production	Advance pay	Salary	Net profits	Profits shared out
Cuba	138	7,559	20,532.2	0.87	8,491.8	2,492.6	3,066.3	2,270.9
Pinar del Rio (dark tob.)	76	3,333	10,206.3	0.91	3,734.1	1,113.5	740.4	859.0
Pinar del Rio (mild tob.)	9	688	1,472.3	1.08	819.6	267.3	−29.4	26.2†
Villa Clara	20	1,536	4,053.2	0.84	1,661.2	409.6	622.7	303.7
Sancti Spiritus	23	1,667	4,321.4	0.70	2,016.5	576.3	1,767.2	1,072.6
Granma	10	335	479.0	1.04	260.4	125.9	−34.6	9.4‡

Source: Compiled with data from ANAP returns for 1985.
*Figures refer to the CPAs for which returns had been computed.
†Figure refers to the three out of the nine CPAs that were profitable.
‡Figure refers to the four out of the 10 CPAs that were profitable.

this juncture, it can be argued that cooperatives provide the structure best suited to rural women's needs in the productive and reproductive processes, helping them out of a subordinate and hence underestimated position in both. While there has been a certain dropping off of female membership, women in cooperatives are now coming forward and questioning more. At the 1982 ANAP Congress, it was cooperative farmwomen delegates who raised the issue of obstacles women face in the cooperatives and called for greater attention on this front. They also highlighted the positive benefits of women's active participation in production and the decision-making process. Their voice at a congress of what has traditionally been a male farmers' organization was both a testimony to change and a challenge for the future.

NOTES

1. Readers are referred to Stubbs (1985).

2. My thoughts on this have been stimulated by the now considerable body of research on Third World, especially Caribbean and Latin American women, the work of Carmen Diana Deere and Magdalena León, in particular, although I have seen little on women in tobacco.

3. Brian Pollitt's work in this respect is interesting. Pollitt (1982) is a more recent overview.

4. Tables for 1945 include Las Villas province, to which Cabaiguán belonged. Under the new 1975 political administration division of the country, Las Villas was divided into several provinces, one of which was Sancti Spíritus, which include Cabaiguán. Tables for the post-1975 period quote Sancti Spíritus province, but there is clearly no point of comparison. Pinar del Río province was left unchanged.

5. A considerable body of information on the two areas can be found in *Cuba Contemporánea* (1944).

6. The history of struggle can be found in Regalado (1979) and Mayo (1980).

7. Pérez (1970) provided much background information, as did Perez (1978a; 1978b)

8. A good overview of agrarian reform can be found in Acosta (1973) and Regueña (1983).

9. Shortage of labor was a constant throughout the 1960s and 1970s, and voluntary temporary brigades would be mobilized for stints in agriculture. In the 1970s, the new schools in the countryside and combined study/work programs were fostered in conjunction with the expansion of the farm sector. In tobacco, student labor is highly seasonal and used for weeding and harvesting.

10. The tobacco growing was particularly dispersed. The agreement reached between private and state sectors was that the latter would take on more of other crops, such as tubers, which required less specialized attention and private tobacco growers would concentrate more on what was their specialized crop.

11. This point comes out clearly in a highly informative work by Adelfo Martín Barrios (1982), himself once a Pinar del Río tobacco grower and now a key ANAP figure. ANAP also has many internal studies of its own.

12. Attention was paid to tobacco as a traditionally important industry and crop, both as a potential foreign exchange earner and major domestic product in a nation of tobacco producers and smokers. In a situation of greater demand than supply, tobacco has been heavily rationed at state-subsidized prices and sold on the free market at much higher prices. The health hazard posed by smoking is an anomaly that is only recently being faced fully, with a major nationwide anti-smoking campaign. The impact this will have on tobacco growing and manufacturing remains to be seen.

13. Tobacco farmers interviewed referred to the disincentives to growing tobacco. They also grew rice, beans, vegetables, etc., which they marketed locally.

14. Policy on this was spelled out in theses and resolutions at the First and Second Party Congresses (1975 and 1980). A good overview of policy and progress is to be found in Fidel Castro's speeches at the close of the Fifth and Sixth ANAP Congresses of 1977 and 1982. Since then, several studies have charted the progress of the cooperative movement. Gómez (1983) is a journalistic account. Trinchet Vera (1984) is the most comprehensive to date, although it has no breakdown by crop or on women. A good introductory cross-country analysis including Cuba can be found in Deere (1984; 1985).

15. Free farmers' markets were legalized in the late 1970s, to sell private farm surplus directly to the consumer. The hope was to boost production and undercut any black market prices. In turn, parallel state markets were introduced, marketing farm surplus on produce in shorter national supply at a higher price to undercut the free market prices, especially as produce from the CPAs became more plentiful. Most CPAs took the eminently political decision to sell only to the state market and not the free market, although they were under no obligation to do so. Free markets were finally eliminated at the request of the cooperatives after the May 1986 Second National Meeting of the CPAs. Tobacco was not such a market crop, rather, it was bought almost exclusively by the state, but both

private growers and cooperatives did grow other staple produce that could be marketed in this way.

16. The economic management and planning system came in the mid-1970s and was an attempt to introduce greater decentralization and flexibility within central planning requirements, as well as a program of material and moral incentives.

17. State farms or enterprises are still as a "higher form" of production. Generally speaking, the rule of thumb is that in areas where there is a need for development plans with high investment, there is a preference for peasant land to pass into the state sector. Where there is no need for high investment but peasant specialization, then the best solution is the union of peasants into cooperatives.

18. Several studies have already charted the progress of the cooperative movement. Gómez (1983) is a journalistic account. Trinchet Vera (1984) is the most comprehensive to date. Neither deals with women.

19. Given that the private sector is an aging sector, cooperatives found substantial numbers of farmers retiring on pensions. There is a case to be made that, with the age factor, migratory and generational patterns will of their own accord disintegrate the peasant economy. This is yet another reason to boost cooperativization and technification.

20. Founded in 1961, the FMC mobilized around job and training opportunities for women. The FMC-ANAP brigades were organized for sporadic and temporary work during the years when many men were mobilized militarily or for harvest periods. By the 1970s, their work was more regularized and the women formed remunerated production brigades. Today the volunteer brigades are almost in abeyance.

21. Participant observation and open-ended interviews were complemented by the compilation of statistical data from the individual state farm, CPA and CCS.

22. Women are employed year round, either in sorting or other stages of tobacco production. If for any reason work is not available, they are guaranteed 40% of salary. CUBATABACO has also invested in improving conditions in the sorting sheds and has upped piece-rates for women's work.

23. Little has been written on rural Cuban women as such, although there is a wide literature on women in general. Readers might refer more particularly to Isabel Larguía and John Dumoulin (1984; 1985). The FMC has detailed Congress documents and reports, working papers and studies of its own, in which reference can be found to the early schools and brigades. A short overview of women in sugar can be found in Pérez Rojas (1986).

24. Women are in the majority (51.6%) in higher agricultural studies for the 20–24 age group.

25. This came out strongly in the media at the time. The press ran features and interviews, and documentaries such as the National Film Institute's 20-min *Tierra sin cerca* (Idelfonso Ramos, 1977) were made. This was also clear from my own interviewing in tobacco areas.

26. This is a national drive. Experienced former workers in crops such as sugar and tobacco are brought in temporarily for the harvest in an attempt to guarantee work productivity. Piece-rates rather than flat rates are also used as an incentive to increase productivity. The smaller plot applies particularly to a crop like tobacco, where each plot is the responsibility of a foreman.

27. This is the balance of money reimbursed to the farmer on sales against repayment of bank loans, as quoted by the Banco Nacional de Cuba (1984).

REFERENCES

Acosta, José, "La revolución agraria en Cuba y el desarrollo económico," *Economía y Desarrollo*, No. 17 (May–June 1973).

Alvarez, Mavis, and J. Stubbs, "La mujer campesina y la cooperativización agraria en Cuba," in Deere and León (Eds.), *La situación de la mujer rural en América Latina y el Caribe y las políticas del estado* (Mexico City: Siglo XXI-ACEP, 1986).

Barrios, Adolfo Martín, *La ANAP, 20 años de trabajo* (Habana: Editora Política, 1982).

Banco Nacional de Cuba, *Crédito al sector campesino y cooperativo* (La Habana, 1984).

Comité Estatal de Estadístico (CEE), *Anuario Estadístico de Cuba* (Havana: 1984).

Concepción Pérez, Rogelio, *Historia de Cabaiguán*, 3 volumes, unpublished (1979).

Conceptión Pérez, R., *Pinar del Río* (Habana: Editorial Oriente, 1978).

Conceptión Pérez, R., *Sancti Spíritus* (Habana: Editorial Oriente, 1978).

Cuba Contemporánea (Havana: Editorial Panamericano, 1944).

Deere, Carmen Diana, "Agrarian reform and the peasantry: The transition to socialism in the Third World," Paper presented at the Seminar on the Problems of Transition in Small Peripheral Economies (Managua, 1984).

Deere, C. D., "Rural women and state policy: The Latin American agrarian reform experience," *World Development*, Vol. 13, No. 9 (1985).

Gómez, Orlando, *De la finca individual a la cooperativa agropecuaria* (Havana, 1983).

Larguía, Isabel and John Dumoulin, *Hacia una ciencia de la liberación de la mujer* (Havana: 1984).

Larguía, I., and J. Domoulin, "La mujer en el desarrollo: Estrategia y experiencia de la Revolución cubana," *Casa de las Américas*, No. 149 (March–April 1985).

Mayo, José, *Dos décadas de lucha contra el latifundismo* (Havana: Editora Politica, 1980).

Minesterio de Agricultura, *Memoria del censo Agricola Nacional*, 1946 (Havana, Republica de Cuba: 1951).

Pérez Rojas, Niurka, "Women in the sugar agroindustry," *Granma Weekly Review* (12 January 1986).

Pollitt, Brian, "The transition to socialist agriculture in Cuba: Some salient features," *IDS Bulletin*, Vol. 13, no. 4 (1982).

Regalado, Antero, *La lucha campesina en Cuba* (Havana: Editora Política, 1979).

Rojas Requeña, Iliana, Mariana Ravenet Ramírez, and Jorge Hernández Martínez, "Desarrollo y relaciones de clases en la estructura agraria en Cuba," in *Estudios sobre la estructura de clases y el desarrollo rural en Cuba* (Havana: University of Havana, 1983).

Stubbs, Jean, *Tobacco on the Periphery: A Case Study in Cuban Labour History, 1860–1958* (London: Cambridge University Press, 1985).

Trinchet Vera, Oscar, *La cooperativa de la tierra en el agro cubano* (Havana: 1984).

·4·

The Performance of the Cuban Sugar Industry, 1981-85

CARL HENRY FEUER

1. INTRODUCTION

It is no longer a handful of men in a small cabin cruiser, armed with ideas rather than with weapons. Now it is an immense and seaworthy ship that no tide, no wind, no storm can cause to founder: a ship laden with many dreams that have become reality and many realities that are still dreams to be realized; a ship where an entire people has set course for the future and lands once again, aware that there may be waves of difficulties and obstacles ahead, or an enemy lying in wait; a ship whose determination, self-confidence and dedication have earned it many more January 1sts. (Castro, 1986, p. 21)

The recent experience of the Cuban sugar industry bears witness to the truth of Fidel's metaphor. We find in that experience ample evidence that many Cuban dreams have become reality: the mechanization of the harvest and the release of thousands of Cuban workers from the hell of cane cutting; the remarkable improvement in the standard of living of sugar workers; the achievement of a stable production rate of about eight million tons of sugar a year; the beginning of the effort to modernize Cuba's antiquated industrial infrastructure; and a fundamental modification of the pattern of unequal exchange which historically characterized Cuba's sugar trade. There are also in that experience, however, many dreams yet to be realized: the failure of Cuba to realize fully its dreams of export-led development due to the abysmal state of the world sugar market; the lack of profitability within the sector; inadequate levels of production of sugar cane; and continued crop uncertainties due to disease and nature. While the Cuban sugar industry has indeed become an "immense and seaworthy ship," a ship that the Cuban people can be proud of, it is still subject to severe buffeting by external forces and internal problems.

In assessing the recent experience of the Cuban sugar industry, we must also bear in mind how important this sector remains within the totality of Cuban political economy. The sugar industry is the source of employment for a large and important segment of the Cuban people, almost 400,000 men and women who, with their families, make up perhaps one-sixth of the Cuban population. Those employed in this sector are not only numerically important; sugarworkers historically have been a backbone of the Revolution. Sugar cane also covers 1.7 million hectares and about one-third of the total means of production used in Cuban industry is located in the sugar industry (*Anuario*, 1983, p. 144). The sugar produced represents 80% of the value of Cuban exports and the industry as a whole contributes 10% to Cuba's Global Social Product (*Anuario*, 1983, p. 881).

A study of the Cuban sugar industry is important also for reasons that extend beyond Cuba itself and its development process. First, Cuba is the third leading producer of sugar cane in the world, the second largest producer of raw sugar, and the world's leader in raw sugar exports. Second, Cuba is a laboratory for studying the role and implications of state management of agriculture in the Third World. In Cuba 80% of the cultivable land is in the state sector, and most of that is land used for the production of sugar cane. No other country in the world has as high a proportion of its land in the state sector. The 153 factories involved in processing the sugar cane, moreover, are also all controlled by the government, via the Ministry of the Sugar Industry (MINAZ).[1]

A third factor that makes studying Cuba's sugar industry important is the rapid strides that Cuba has made in mechanizing its sugar-cane harvest and in modernizing its sugar sector, without wholesale layoffs or the marginalization of the work force. By the end of the 1970s, according to one scholar, Cuba's sugar economy "was increasingly assuming characteristics — expressed in terms of equipment, technique, labour skills and productivity trends — more commonly associated in the Third World not with agricultural activity but with industry" (Pollitt, 1981, p. 8). This is a major achievement for a Third World country.

In this paper I will try to summarize the recent experience of the Cuban sugar industry, focusing especially on structural changes, production, patterns of investment, conditions for the workers and exports.

2. ORGANIZATION OF SUGAR PRODUCTION: STRUCTURE AND CHANGE

Practically all the sugar cane grown in Cuba today comes from either state farms or cane cooperatives, with four-fifths of the total grown on the state farms. All of the cane is processed in state-owned and managed factories; transportation, storage, marketing and input provision are also government functions. Two-thirds of the cane is cut by machine; the rest by professional, paramilitary and volunteer cane cutters. While this picture of socialized, state-run production has remained in broad outline the same since the early 1960s, the organizational structure of the industry did change significantly during the first half of the 1980s. The government integrated the operations of the state farms and factories, forming huge agro-industrial complexes; modified the activities and responsibilities of the work brigades, in keeping with the New System of Economic Management and Planning (SDPE), adopted a policy of self provisioning for the state farms; and oversaw the voluntary and rapid amalgamation of *campesino* farms into cane cooperatives.

(a) *The agro-industrial complexes*

The formation of the agro-industrial complexes began in 1980, when MINAZ assumed responsibility for the agricultural side of the industry, the cultivation, growing, reaping and transportation of the sugar cane. Prior to this, MINAZ had responsibility only for milling and marketing operations, with the Ministry of Agriculture responsible for the rest. This step toward integration freed up the Agriculture Ministry to concentrate on food production, and facilitated the establishment of centralized field-factory complexes in the sugar industry (Benjamin, 1984, p. 146).

The first four such administrative units linking, in each case, a factory with the state farms supplying it with sugar cane, were established in preparation for the 1981 crop. Seventeen were in operation in 1982, 92 in 1983 and, by 1986, every factory had become administratively linked to the farms producing its cane. The cane cooperatives, however, maintained their administrative independence (Pérez, 1985, p. 2; *Latin America Commodity Reports*, 11/12/82, pp. 6–7).

(b) *Permanent production brigades*

The 1980s was also a period of experimentation with and implementation of "new type," "permanent" or "integrated" brigades. This was a type of decentralization of decision making and responsibility to the grassroots production units. The brigades were assigned their means of production and could make decisions about how best to utilize the equipment, raw materials, energy resources, wages and other resources at their disposal in fulfilling their production plan. They received bonuses based on their performance and cost savings. By 1985, there were 104 such brigades of workers operating in 17 sugar agro-industrial complexes (10% of the total number of permanent brigades in the agricultural sector) (*Granma Weekly Review*, 2/9/86, p. 4). The brigades encompassed almost 21,000 sugar workers; 78 of the brigades were working in the canefields, 18 were in the factories and eight worked in the area of self provisioning (*Granma*, 9/26/85, p. 3; 12/13/85, p. 3). The early results of the canecutter brigades were impressive (Table 1).

This process of decentralization of authority to the workers will continue in the future. The plan

Table 1. *New canecutter brigades, plan fulfillment, first half 1985*

Index	Performance
Gross production	105% of plan
Wage fund	96% of plan
Productivity	109% of plan
Cost of materials per peso of gross production	reduced by 9 cents
Wage cost per peso of gross production	reduced by 4 cents

Source: *Granma Weekly Review* (2/9/86), p. 4.

for 1986–90 calls for a 3.5% average annual increase in labor productivity, and links this to "better organization of work based on a broader use of permanent and comprehensive brigades and cost-profit accounting at the brigade level" (Castro, 1986, p. 8).

(c) *Self provisioning*

An increased emphasis on self provisioning — the allocation of land and work time by the state farms to the production of food for the workers — also characterized the first half of the 1980s. This was not a return to peasant private plots, but state farm land set aside for collective production. The produce was served in the workers' cafeterias or sold to the workers at low prices. After a few years of experimentation, the authorities were so pleased with this innovation that in 1983 they decided to extend it to all state farms, from the one-half that had such plots at that time (Benjamin, 1984, p. 45).

The cane farms apparently moved the quickest in this direction. Several rapidly achieved self-sufficiency in meat and milk; overall, by 1983, sugar workers provided 40% of their meals in this way. The cane farms were perhaps most advanced because of the availability of valuable byproducts of the production process that could be used for animal feed (Benjamin, 1984, p. 147).

All, however, was not roses. There were abuses:

> It led to increased, sometimes excessive, supplies of goods which those workers also received through other channels. Lands were used to these ends, to the detriment of State enterprises' basic production. (Castro, 1986, p. 6)

In other words, self provisioning was expanded, in some cases, to the provisioning of others through sales on farmers' markets or in other ways. The abolition of the farmers' markets in 1986 was probably aimed, in part, at ending this problem.

(d) *Cooperatives*

The fourth significant organizational change affecting the sugar industry in recent years has been the voluntary amalgamation of independent campesino cane producers into producer cooperatives. This movement harkens back to the first agrarian reform when there was a brief period in which the expropriated cane farms were established as cooperatives. Shortly, however, these cooperatives, which in fact were always dominated by the state, were officially transformed into *granjas*, or state farms. For the first 15 years or so of the Revolution, then, sugar cane was produced only on state farms and independent campesino farms, with the latter producing between one-fifth and one-quarter of the total.[2]

In the mid-1970s, this situation began to change. At the First Party Congress in 1975, Fidel talked about the development of peasant cooperatives, seeing them as part of the transition of the peasantry to "superior" forms of production. The Fifth National Conference of ANAP in 1977 took this one step further, establishing the promotion of Agricultural and Livestock Producer Cooperatives (CPAs) as a priority, and approving guidelines for their formation and organization as self-managed entities (Deere, 1984a, p. 57).[3]

In the ensuing eight years, Cuban peasants formed CPAs at a rapid pace (Table 2). By 1985, about two-thirds of all the private land was in cooperatives, compared to 11% in 1980. The cooperatives included over 71,000 campesinos as members (*Granma*, 10/12/85, p. 2; Castro, 1981, p. 64).[4]

The development of cooperatives was "most intense in the cane sector" (Pérez, 1985, p. 2). By 1985 there were about 430 cane cooperatives encompassing 400,000 hectares for an average cooperative size of 930 hectares.[5] Over 31,000 campesinos were members and they produced about two-thirds of all the cane originating in the private sector (*Granma*, 3/18/85, p. 2; 10/12/85, p. 2; 10/23/85, p. 2; Ramírez, 1985, p. 20). The cooperatives also apparently achieved a level of mechanization, at least in terms of the harvest, that was equivalent to the state farms. In the 1985 crop, 485 combines reaped 5.3 million metric tons of cane on the cooperatives, or about 65% of their total production, the same percentage as on the state farms (*Granma*, 10/12/85, p. 2 and 10/23/85, p. 2).

Table 2. *Development of producer cooperatives in Cuba*

Year	Number of CPAs	Total land (hectares)	Average size (hectares)
1974	43	n.a.	n.a.
1978	363	16,721	46
1980	1,017	197,274	194
1985	1,378	1,008,755	732

Source: Castro (1981, p. 64; 1986, p. 11).

While Cuban officials are committed to the state farms, the formation of the cooperatives provides for the first time a meaningful, more participatory, alternative to state management. Along with other factors, such as the innovative new type of production brigades, the emphasis on self provisioning, the new incentive package (see below) and the advances by women in this sector of the economy (see below), cooperativization reflected an important trend toward greater worker empowerment and democratization.

3. PERFORMANCE 1981–85

(a) *Production*

Between 1981 and 1985 the production of sugar increased 12% over the previous five-year period, and a relatively stable level of eight million tons per year was achieved (Castro, 1986, p. 3). While this continued the pattern established in the 1970s of secular growth and development in the sugar sector (Table 4), it was in other respects a disappointing performance. It was only about half the growth originally targeted and much less than the growth rate registered in the previous five-year period (Castro, 1981, pp. 16–17, 42). It also did not match "the possibilities created by invested resources," and was insufficient to meet Cuba's critical need for foreign exchange (Castro, 1986, p. 6).

Yields per hectare also continued to increase steadily, in a pattern established in the 1970s. As in the area of production, however, the rate of growth in yields — 9% over the five years — was down compared to the previous quinquennium.

Before going further, it is important to put this performance into context. Only two of the seven leading producers in Latin America and the Caribbean showed a higher rate of growth of sugar production than Cuba in this period (International Sugar Organization, 1984). Using another indicator, FAO's more broadly based Crop Production Index, Cuba had the highest rate of growth between 1976–80 and 1981–85 of all 30 countries in the Americas (*FAO*, November 1985). This was an exceptional achievement, especially given Cuba's relative degree of income equality.

Another important achievement during this period was the increase in the productivity of cane cutters and other workers. In 1985, the *macheteros* harvested a daily average of 3.78 tons, an increase of 36% over the figures for 1981. According to Fidel this was due to "the mobilization of more productive sugar-cane cutters and better organized brigades"; as a result fewer volunteer cutters were needed (Castro, 1986, p. 11). There were other manifestations of improved work efforts. The workers

Table 3. *Growth of the cooperative sector, 1983 to 1986*

Date	Number of coops	Total land (hectares)	Average size (hectares)	Members
1983	1,480	777,947	526	78,000
1985	1,410	997,898	707	72,564
Oct 1985	1,393	1,001,575	719	72,246
Feb 1986	1,378	1,008,755	732	n.a.

Source: Benjamin (1984, p. 167); Ramírez (1985, p. 20); *Granma*, 10/12/85, p. 2; Castro (1986, p. 11).

Table 4. *Cuban sugar performance, 1957 to 1985*

	1957	1962–70 (average)	1971–75 (average)	1976–80 (average)	1981–85 (average)
Raw sugar output (metric tons, '000; crop year)	5,742	5,370	5,548	6,930	7,776*
Industrial yield (tons sugar extracted, 100 tons cane)	12.80	12.04	11.38	11.22	10.87†
Sugarcane harvested (metric tons, '000)	44,700	50,326‡	49,495	65,016	71,480*
Area harvested (hectares, '000)	1,265	1,077‡	1,158	1,261	1,245†
Yield (ton/ha)	35.0	46.7	42.7	51.6	56.1†
Mechanized harvest (%)	0	0	13	39	54
Campesino production (% of total)	n.a.	26.7	19.1	17.9	18.6
Yield, campesino sector (tons/hectare)	n.a.	47.4	45.5	57.9	59.6
Yield, state sector (tons/hectare)	n.a.	45.5	42.2	50.4	54.8
Harvest length (days)	88§	n.a.	n.a.	154	149†

Source: National Foreign Assessment Center (1981, p. 3); *Anuario* (various years); FAO (January 1986, p. 1); Edquist (1985, p. 38); Gómez (1985, p. 2); Varela (1985, p. 3).
*Figures for 1984 and 1985 are unofficial.
†1981–83 only.
‡1965–70 only.
§1954 crop.

reduced the amount of extraneous matter — cane tops, straw, stones and earth — taken up with the cane from 6.47% in 1984 to 5.97% the following year and there was a reduction in 1985 of cane losses, from cane left in the fields, of over three million tons (*Granma*, 9/5/85, p. 1).[6]

(b) *Mechanization*

The mechanization of the harvest proceeded rapidly during the 1980s. In 1980, only 45% of the crop was cut mechanically; by 1985, almost two-thirds of the crop was cut in this way (62%). By the mid-1980s, over 4,000 cane combines were in operation and the Holguin facility was producing the units at the rate of about 600 per year. In addition, in 1986, Cuba began producing a more powerful combine, the KTP-2, and started experiments with a newer unit, the KTP-3 (Castro, 1986, p. 3; *Granma Weekly Review*, 3/30/86, p. 7). The mechanization of the harvest complemented Cuba's earlier success in completely mechanizing the loading of the cut cane. Prior to 1959, workers both cut and loaded the cane manually. By the mid-1970s, the loading operation was completely mechanized; Cuba is now well on its way to duplicating this achievement for the harvest.[7]

Productivity, however, did not keep up with the process of mechanization. In 1985, 85 tons were cut per machine per day, down from 90 in 1981 and slightly higher than the figure for 1980. This compared to the standard of 100 tons per day set by the sugar authorities. The rapid introduction of cane combines, then, was not without its difficulties — low qualifications of operators, repair difficulties, organizational deficiencies, and lack of stability in the work force (Varela, 1985, p. 3; *Granma*, 7/16/82, p. 3).

Whatever costs and problems Cuba may have experienced, mechanization was extremely beneficial. It resulted in:

(1) the transferrence to thousands of workers of the skills to operate and repair complex farm machinery;
(2) the freeing up of the labor power of around 200,000 Cuban workers, no longer needed in the harvest, resulting in a substantial decline in the percentage of the labor force engaged in agriculture;
(3) the establishment of a nationwide system for the maintenance and repair of harvesters;
(4) the acquisition by many Cuban technicians of valuable design experience, and experience in transforming design into an actual manufacturing process;

(5) the development of a new industry, involving the manufacture of cane harvesters;

(6) the development of horizontal linkages within Cuba — adjoining the harvester factory, for example, a plant producing tractor implements, and employing thousands of workers, has been built.[8]

Mechanization also facilitated the reintroduction of green-cane harvesting in Cuba. Burning the cane prior to harvesting was introduced in Cuba in 1971. By 1972, almost three-quarters of the canefields were burnt. One of the major reasons for this was that it increased the productivity of cane cutters, thus reducing the labor scarcity problem. Another reason was that the first mechanical harvesters worked better in burnt cane. But burning also has negative features. It reduces the fertility of the fields and leads to rapid deterioration of the cut cane. The sucrose content is reduced unless the cane is processed quickly. Harvesting unburnt, green cane, is much preferred. Improvements in mechanization made possible the reintroduction of green-cane harvesting in Cuba in the late 1970s; by 1985 about four-fifths of all the cane was cut green, with burning almost solely limited to areas set aside for replanting (Edquist, 1985, p. 116; Varela, 1985, p. 3).

(c) *Energy efficiency*

During the first half of the 1980s, in another major achievement, the sugar industry moved rapidly to modify its pattern of energy use. The consumption of oil, which averaged one gallon/ton of cane processed in 1980, was totally eliminated by 1986, saving 1.7 million tons of oil. For the most part, bagasse was substituted for the oil. As a result, the sugar industry was the major contributor to Cuba's annual petroleum savings of almost two million tons between 1981 and 1985. This was of great significance to the Cuban economy since any saving on its quota of oil from the Soviet Union was re-exported, resulting in significant foreign exchange earnings (Castro, 1981, p. 17; 1986, p. 3; *Mundo Azucarero*, September 1985; *Granma*, 10/17/85, p. 1).

(d) *Production problems*

Growth in production and yields came despite significant setbacks, many of them uncontrollable, and showed Cuba's continued ability to "confound its critics" (*Latin America Commodities Report*, 6/8/79, p. 85). The 1981 crop, coming on the heels of an outbreak of rust disease affecting upwards of 40% of Cuba's sugar cane, was one example. Despite Western predictions that the crop that year would not even reach six million tons of sugar, due to the lingering effects of the disease, the immense effort needed for replanting, unseasonable rains, and other factors, production actually exceeded 7.3 million tons (Table 5) (*Latin America Commodities Report*, 10/24/80, p. 1 and 11/21/80, p. 4). Cuba could "confound its critics" because it had achieved a high level of organization and modernization in the industry, and because of the continued success of its mobilization efforts.

While Cuba could, at times, rise to the occasion, the overall performance of the industry was still disappointing. Weather was certainly one factor which contributed to production being below expectations. During three of the five years rainfall was less than 80% of the historical average during the rainy season; the 1985 rainy season was the driest since 1959. Drought was complemented by unseasonable torrential downpours, such as that which hit the sugar harvest during the first quarter of 1983, and by five tropical storms during the period. The rain which hit early in 1983 was the worst in 20 years, interrupting what had started off as an excellent crop (Castro, 1986, p. 8; *Latin America Commodities Report*, 2/25/83, p. 5). A serious and prolonged drought during 1985 convinced Cuban planners that the 1986 crop would be reduced by as much as one million tons. As a result, in July 1985, Cuba was forced to purchase half a million tons of sugar on the world market, to ensure that future delivery commitments to the USSR would be fulfilled (*Latin America Commodities Report*, 10/11/85, p. 8). Cuba's weather woes were capped off by Hurricane Kate late in 1985. Almost one million hectares of cane were damaged; flooding also wiped out some crude sugar being stored in warehouses. This was a major catastrophe that severely set back Cuban plans for the 1986–90 period (*Granma*, 11/22/85, p. 1).

Another factor which was disconcerting was the poor performance of the campesino cane producers during 1984 and 1985. Despite the effort Cuba has made to increase the production of sugar cane, mainly through increased yields, campesino production in 1985 was significantly below what it was in 1979, with yields down 16%, to 54.3 tons per hectare (Table 6). For the whole 1981–85 period the average yield was 59.1 tons per hectare, only marginally greater than the average for the previous quinquennium (Table 4). Historically the campesinos have outperformed the state sector, averaging 15% higher yields during the 1976–80 period.[9] This made the

Table 5. *Sugar production statistics*, 1981 *to* 1985 (Crop years)

	1981	1982	1983	1984	1985
Sugarcane harvested (metric tons, '000)	66,600	73,100	69,700	75,000*	73,000*
Area harvested (hectares, '000)	1,209	1,327	1,200	n.a.	n.a.
Yield (tons/hectare)	55.1	55.1	58.0	n.a.	n.a.
Raw sugar output (metric tons, '000)	7,359	8,210	7,109	8,200*	8,004*
Industrial yield (tons sugar extracted, 100 tons cane)	11.08	11.17	10.35	n.a.	n.a.
Sugarcane harvested, state sector (tons, '000)	54,500	60,200	54,900	60,798*	60,626*
Yield, state sector (tons/hectare)	53.8	53.9	56.7	n.a.	n.a.
Sugarcane harvested, campesino sector (tons, '000)	12,100	12,900	14,800	14,202	12,374
Yield, campesino sector (tons/hectare)	61.3	61.0	63.6	57.7	54.3

Source: *Anuario* (1983); Gómez (1985, p. 2); López (1986, p. 4); *Latin America Commodities Report* (8/3/84).
*Unofficial.

Table 6. *Campesino cane production, 1979 compared to 1985*

	1979	1985
Production (metric tons, '000,000)	13.51	12.37
Area harvested (hectares, '000)	208.9	227.8
Yield	64.6	54.3

Source: *Anuario* (1981, p. 87); Gómez (1985, p. 2).

decline in campesino production especially disturbing: it roughly corresponded to the growth of cane cooperatives, which were responsible for about two-thirds of the campesinos' cane output. In addition, the government had allocated considerable capital to the cooperatives for the purchase of machinery and equipment (*Latin America Commodities Report*, 11/12/82, p. 7).

It appeared that a combination of factors conspired to reduce the production and yields of the campesinos' canefields after 1983: low prices for the cane from the state; the possibility of marketing alternative crops at the new farmers' markets; poor management — wasting investment, not planning tasks properly, not using machinery and transport equipment well — leading to higher costs of production; low level of technical skill; and difficulties with the availability of spare parts. It is likely that the farmers had difficulty coping with the newly formed, large, cooperative farms. The bottom line was that the canefields were not getting the attention the government thought they needed (Batista *et al.*, 1985, p. 11; Machado, 1986, p. 5; *Granma*, 9/23/85, p. 3; Gómez, 1985, p. 2). In late 1985 the government introduced a new price system in an effort to cope with the problems of the farmers. Starting with the 1986 crop, preferential prices were paid to farmers who achieved yields higher than they averaged during the three previous crops, as long as they maintained at least the same amount of land in cane as was the case in 1984. Other measures, including a government subsidy to campesinos for putting new land into production, were also adopted (*ANAP*, 1986; Gómez, 1985, p. 2). In 1986, moreover, the farmers' markets were closed, reducing the campesinos' outlets for alternative crops.

There were other factors besides disease, weather and disappointing efforts by the cooperatives and private farms which prevented higher production and yields from being achieved:

> Sugarcane production targets were not met because of insufficient planting and unsatisfactory agricultural yields; inadequate soil preparation, short supplies of agricultural machinery drawn by high-power tractors and improper field leveling and drainage resulted in high crop losses. Delays in the development and introduction of new, more

productive and disease-resistant varieties; low-quality seeds; ill-timed planting; weeding and cultivation delays, among other factors contributed to the low density of many cane fields, which caused limited agricultural yields. (Castro, 1986, p. 6)

There is some data to back up Fidel's case. Between 1981 and 1984, for example, Cuba was beset by a continuing inability to fulfill replanting targets (*Granma*, 1/8/85). The average annual amount of land planted with new seed cane during this period was down 6% from the 1976–80 period, despite the record planting in 1981 to replace cane affected by rust disease (Table 7). This no doubt had an effect on yields, since older, or ratoon, cane characteristically yields less than young cane.[10]

The scarcity of foreign exchange also forced Cuba to economize on key inputs. In particular, the importation and use of herbicides was cut back during the 1980s while the application of fertilizer showed only a minimal increase (Table 7). Fertilizer utilization was also affected by transportation problems and losses due to inferior packing, handling and storing (Castro, 1986, p. 6). Because of the cutback in the use of herbicides, Cuba had to mobilize twice as many workers as usual — 250,000 — to weed the sugar cane. This mobilization proved difficult, especially when the harvest extended for a long period. As a result, there were frequent reports in *Granma* of difficulties in this area (*Latin America Commodities Report*, 5/21/82, pp. 2–3; 7/16/82, p. 3). The failure to extend substantially the irrigated area was also noteworthy. In 1983, only 346,000 hectares had irrigation, a scant 5% more than in 1980 (*Anuario*, 1983, p. 195; 1981, p. 86).

Yet the Five-Year Plan for sugar called for a 75% increase in the total area to be irrigated (Castro, 1981, p. 42).

4. INVESTMENT

Investment in the sugar industry, primarily in processing operations, was pushed forward in the 1980s at a substantial rate. Between 1980 and 1986, seven new mills were built in Cuba, and an eighth was constructed in Nicaragua. These were the first new mills constructed in Cuba in over 50 years. Two others were under construction in 1986. In addition, 38 existing mills were enlarged and/or modernized. All this resulted in a substantial increase in daily sugar-cane grinding capacity of more than 80,000 tons. In addition to reflecting the continued commitment of the leadership to the sugar industry, this capital development program also manifests Cuba's technological progress.[11] The new mills were designed by Cuban technicians and between 45% and 60% of the equipment was built in Cuba; it should not be surprising that the machine industry in the sugar sector doubled its production between 1981 and 1985 (Castro, 1981, p. 16; 1986, p. 3; *Granma Weekly Review*, 5/11/86). A substantial amount of the cost of this building program was provided by CMEA. In 1981, CMEA provided a grant of US $643 million in cash and materials to Cuba for the construction of new mills. Most of this funding came from the USSR, East Germany and Bulgaria (*Latin America Commodities Report*, 8/28/81, p. 1).[12]

These and other investments in the sugar

Table 7. *Summary of agricultural indicators, 1971 to 1983*

	1971–75 (ann. average)	1976–80 (ann. average)	1981–83 (ann. average)
Nitrogen fertilizer (kg/ha)	158	185	194*
Herbicides (area getting at least one application, hectares, '000,000)		1.09	1.07*
Irrigated area (hectares, '000)	169.1	285.3	337.3
Irrigated area as a percent of total area in cane	11.7	17.5	19.4
Area sown with new cane (hectares, '000)	390.9†	375.9	352.1‡
Area sown as a percent of area in cane	27.1†	23.1	21.5

Source: CEE (1982, Table 4.6); *Anuario* (1983); *Anuario* (1981); *Granma* (1/8/85).
*1981 and 1982, average.
†1972–75 only.
‡1981–84.

industry — for example, the construction of 239 new sugar-cane receiving and cleaning centers between 1981 and 1985 — reflected Cuba's view that industrialization should "become the most important factor in our country's economic development" (Castro, 1986, p. 3). As a result, most of the investment seemed to be in the industrial side, rather than agriculture. Between 1976 and 1980, for example, overall investment in agriculture (including both sugar-cane and non-sugar-cane sectors) sank to only 19% of total investment, compared to 40% in 1971–75 (Castro, 1981, p. 16). The danger in such a pattern is that it could produce a growing disjuncture between the agricultural and processing sides of the industry. If this occurs the production of sugar cane will increasingly become a limiting factor in the further growth of sugar production. There was some evidence between 1981 and 1985 that this was indeed happening, as I have described above.

Policy changes announced late in 1984 suggest a continuation, if not exacerbation, of this trend. These changes reflected Cuba's critical foreign exchange needs and stagnation on world sugar markets. At the same time, Cuba's tendency to transfer a portion of its CMEA sugar quota to the world market (for convertible currency), when production declined, had apparently come under fire. The new policy for investment suggested that Cuba would emphasize the development of sugar byproducts, the export of technical skills and, perhaps, cane combines and rum. Investment in the industry proper, an industry which perhaps had reached its limit with respect to energy saving and convertible currency earnings, would thus probably decline (Fuller, 1985, p. 11). The Plan for 1986–1990 confirmed this development.

5. LABOR

About 360,000 people work in the Cuban sugar industry, roughly 12% of the labor force. Two-thirds of the total are field workers, including cane cutters and machine operators, about 100,000 work in the factories, and there are 30,000–35,000 campesinos actively involved in the production of sugar cane (*Granma*, 3/14/85; *Anuario*, 1983, p. 141; Ramírez, 1985, p. 20; FAO, 1985, p. 66).

While Cuban women have made significant gains within the sugar industry, their position still lags considerably behind their integration into other sectors of the economy. In 1985, women comprised 13.6% of the labor force in the sugar factories, 17.9% in the agricultural operations of the state farms, and 18% of the membership of the cane cooperatives (Pérez, 1986, p. 2). Seventeen percent of the delegates to the 17th Congress of the National Union of Sugar Workers in December 1985 were women (*Granma*, 12/12/85, p. 1). This should be viewed, however, with the knowledge that before the Revolution women were rarely to be found working in the sugar sector. By contrast, 37% of the overall Cuban work force is comprised of women, in addition to 27% of the work force on all cooperatives (Pérez, 1986, p. 2).

It is likely that the percentage of women working in the sugar agroindustrial complexes, particularly in technical positions, will increase in the future. This is due to the fact that women are enrolled in large numbers in technical and university-level courses related to sugar production. During 1984–85, for example, over 4,500 women were enrolled in a variety of engineering, chemistry, sugar production and machine building courses related to the industry. Perhaps portending the level of integration of women into the sugar work force in the future, women in 1985 accounted for 37% of the staff at the sugar industry's four research centers and 49% of all quality control personnel (Pérez, 1986, p. 2).

The heavily ingrained machismo of the sugar industry, however, is reflected in the continuing strength of opposition to women working, and in the retention of restrictions on the kinds of work women may do. While women get equal pay for equal work, the difficulties inherent in overcoming the domestic division of labor mean that women work fewer days and/or fewer hours than men, hence receive less pay. Labor oversupply problems also work against the full incorporation of women into the sugar industry work force (Pérez, 1986, p. 2). Finally, if *Granma* coverage of the sugar industry is any indicator, women still have a long way to go. While *Granma* provides extensive coverage of the industry, especially stories about workers, there is hardly ever any mention of a woman.[13]

A group of sugar workers that *Granma* neglects as much, if not more, than women are the field workers, those involved in weeding, fertilizing, irrigating and spraying the fields. I would estimate that this category encompasses over 100,000 workers, more than one-third of whom are women. Not only is this the largest single category in the sugar work force but by far the area in which women are most heavily concentrated.[14] While *Granma* reports many stories about the care of the cane, including weeding and replanting, few of these describe workers, whether individuals or brigades. This is in marked contrast to its reportage of factory workers, cane cutters and combine operators.

With the mechanization of the harvest, the number of cane cutters has dropped sharply over the years. There were 72,000 cane cutters in the 1985 crop, a reduction by half since only 1980. There were 350,000 cutters in the 1960s (Castro, 1986, p. 3; *Granma*, 9/21/85). The cane cutters include voluntary workers from the labor unions outside of sugar who work for the entire harvest and often successive harvests; campesinos organized by ANAP; militarized cutters from the Ejercito Juvenil del Trabajo, a paramilitary body founded in 1973; and regular cane cutters. In 1981 about 50% of the macheteros were not "professional" cane cutters (*Latin America Commodities Report*, 12/18/81, p. 4).[15] The productivity of the cane cutters is high, reaching a record in 1985 of 3.76 tons per cutter per day. The increasing productivity is a result of Cuban efforts to link material incentives to productivity, as well as the increasing professionalism and skill of the remaining cane cutters.

Sugar workers are not only an important economic force in Cuba, but they also play an important political role. Sugar workers have been a backbone of the Revolution, as they have been historically within the trade union movement. They have been among the most militant of Cuban workers. In three factories described by *Granma*, for example, 18% of the workers were members of the Communist Party and 8.5% were members of the Communist Party youth wing (*Granma*, 5/20/85; 3/14/85, p. 2 and 4/1/85).[16]

Sugar workers received a wage increase in 1980 along with other workers; their basic wages, moreover, are 15% higher than those of people doing comparable work in other sectors. In addition, sugar workers benefited from the following measures instituted as part of the 1980 reform package: a system of bonuses for night work; a wage increment based on seniority for mill workers and combine operators; special incentives for good quality work and for good attendance (Castro, 1981, p. 17).[17] The introduction of a fourth shift in the factories, allowing workers for the first time to take a weekly day off during the harvest, was particularly welcomed. It was said to have had a "dramatic effect" on daily attendance, though it also necessitated the employment of more workers (*Latin America Commodities Report*, 12/18/81, p. 5 and 11/12/82, pp. 6–7). There is also extra pay for older sugar workers, as an incentive not to retire, and for those doing particularly strenuous jobs. Finally, exemplary sugar workers also have greater access to scarce goods, such as appliances, motorcycles, cars and vacations.[18] As a result of these measures, factory workers in the industry averaged 230 pesos per month in 1982, 35% higher than industrial workers in other sectors; field workers averaged 186 pesos per month, 37% higher than other agricultural workers (Benjamin et al., 1984, p. 173).[19] Between 1980 and 1982, sugar-worker monthly wages increased 40% (*Latin America Commodities Report*, 12/18/81, p. 4).

The changes initiated in the 1970s to empower the workers, through the unions, to play a more significant role in the workplace, continued in the first half of the 1980s. Both the Production Assemblies — general meetings of the workers at each enterprise occurring at least every two months with the power to make non-binding recommendations to management on matters affecting work and performance — and the Management Councils — composed of elected workers, representatives of the PCC and Women's Work Front, and management, but again only with the power to discuss and make recommendations on administrative matters — continued to function. The leadership, however, was not satisfied with the progress made in institutionalizing worker participation. In his Main Report to the Third Party Congress, Fidel criticized the failure, until "very recently," to effectively involve the workers in planning (Castro, 1986, p. 7). At the same time, there was an increase in the powers accorded management *vis-à-vis* workers. Management was granted the power to suspend or dismiss workers without having to request authorization from Workers Councils, composed entirely of workers elected to deal with all labor grievances. The workers could only appeal through the formal judicial system (Benjamin, 1984, p. 172; *Latin America Regional Report, Caribbean*, 2/19/82, p. 3).[20]

6. EXPORTS

The development of the Cuban sugar industry since the early 1960s has been strongly affected by the growth of trade relations with the Soviet Union. Prior to the establishment of these ties, "sustained planned expansion not only of the sugar sector but of the economy as a whole was quite impossible" (Pollitt, 1981, p. 3). While the Soviets could neither alter low and fluctuating world market prices nor act to expand Cuba's free market sales, they did provide a new source of support for Cuban sugar sales. Cuba obtained large markets, long-term agreements, relatively stable and remunerative prices linked to the cost to Cuba of key imports, and, at least for a time, greater flexibility in the allocation of its sugar exports (the Soviet share of Cuban sugar exports falling as world prices rose) (Pollitt, 1981, p. 3).

Linking the price of Cuban sugar to import prices represented a radical change in the pattern of global trade relations.

During the first half of the 1980s, the Soviet Union continued to be Cuba's main customer, taking on average half of all sugar exports; with CMEA the total came to almost two-thirds (Table 8). China, always Cuba's second best customer, increased its sugar imports from Cuba considerably. On average it imported almost as much as all of CMEA, excluding the Soviet Union. In 1985, Cuba and China reached an agreement for China to buy 800,000 tons annually, 8% above its average annual purchases between 1981 and 1984, and double its earlier rate. China's vast market represents an important opportunity for Cuba.

initiatives to advance the socialization of agricultural production have been oriented toward the establishment of large, state-managed agricultural enterprises. These were seen as a "higher stage" in the development of the agricultural sector. This choice of development path, with the sugar industry as a prime example, is controversial. State enterprises are frequently criticized for being bureaucratic, hierarchical and insufficiently democratic in their operations — too much like the capitalist enterprises they replace. Another line of argument, the favorite of capitalist ideologues, is that such enterprises are inherently inefficient, or less efficient than privately-run establishments in a more market-oriented economy. There is evidence to support both lines of argument.

Table 8. *Sugar exports*, 1971 to 1984

	1971–75	1976–80	1981–84
Total exports (metric tons)	5,136,550	6,538,689	7,153,583
Share going to USSR (percent)	36.3	53.0	50.8
Share to USSR & CMEA (percent)	52.6	64.3	64.9
Share to free market (percent)	37.6	27.9	24.7
Share to China (percent)	6.2	6.2	10.4
ISA daily price, average (cents/lb)	14.3	13.2	9.7

Source: International Sugar Organization (1977; 1980; 1984).

Cuba's free market sales dropped substantially during the 1980s. In 1984, Cuba sold about one-third less sugar on the free market than it had in 1979, a decline of more than 600,000 tons, worth about $73 million in 1984 (International Sugar Organization, 1984). Since sugar sales represent a major source of Cuba's hard currency earnings, the sharp drop in world prices, along with the cut in sales on the free market, was a significant financial blow.

7. CONCLUSION

I would like to conclude by trying to integrate the preceding summary of developments within the sugar sector during the first half of the 1980s into a brief discussion of Cuba's development options and choices.

(a) *State farming*

From the early days of the Revolution, Cuban

While recent evidence for Cuba is fragmentary, it provides little sustenance for those who oppose state enterprises on economic grounds. During the first half of the 1980s, the sugar industry performed admirably, if below needs and expectations. Production and yields continued to grow, though at a diminished rate. The growth was substantial enough, however, to allow Cuba to maintain a preeminent position among agricultural producers in Latin America and the Caribbean. But what was the cost of this growth? There is little available information on the costs of sugar production in Cuba.

Cuba's experience with cane cooperatives, a more democratic mode of organizing production, may have strengthened the position of those who see this as a clearly inferior option compared to state farms. As reported above, there was a sharp drop in yields and production in the campesino sector after 1983. It is likely, but uncertain, that problems of management and incentives with the new cooperatives were involved. A new price structure, providing incentives keyed to increasing yields, was one government response; the dissolution of the farmers' markets may have been another.

(b) *Sugar-cane monoculture*

Should Cuba have focused its development so heavily on sugar-cane monoculture? In some respects, Cuba's recent experience provides ammunition for those critical of its heavy emphasis on sugar cane, to the detriment of other crops. The continuing stagnation on the world market, even though cushioned somewhat by CMEA trade arrangements, makes this development choice look wrong. The recent plunge in world oil prices is likely, moreover, to "soften" the CMEA cushion. The 1980s also reinforced the view that sugar cane is particularly susceptible to natural calamities, whether disease, hurricane, or unseasonable rain, and therefore an ill-suited basket in which to place all of a country's eggs.[21] On the other hand, what is becoming increasingly obvious is that Cuba has achieved a very powerful market position and is building a thoroughly modern industrial infrastructure for the production of sugar. This not only puts Cuba in an enviable position when market conditions improve, but also represents important developmental implications, as I briefly suggested above in the discussion on harvest mechanization.

Cuba is also using its advances in this area to diversify production. Cuba is today a world leader in the development and use of derivatives or byproducts from the sugar production process. There are now 10 plants, for example, which produce paper, newsprint, cardboard and particle board from bagasse and 11 plants to manufacture torula yeast from sugar and urea. The yeast is mixed with molasses to yield a high protein animal feed which is cheaper than corn and soybean feeds. By 1985, in addition, almost one-fifth of the sugar centrals were using a Cuban-developed system for recycling factory wastes for use as fertilizer (Castañeda, 1986, p. 12).

(c) *Labor conditions and productivity*

Another major debate surrounds Cuba's effort to maximize social equality and the benefits accruing to the work force, while increasing worker productivity and the workers' identification with the needs of the economy. During the first half of the 1980s, Cuban sugar workers experienced a significant improvement in the quality of their lives, measured by food availability, purchasing power, housing construction and reform, and continued Cuban progress in health care and education. Sugar workers continued to enjoy higher levels of income than many of their Cuban comrades and there were no mass layoffs as occurred in many other countries which produce sugar cane. The emphasis on material incentives which began in the 1970s and was reinforced more recently was linked to an increase in labor productivity. Between 1981 and 1985, for example, the productivity of cane cutters increased 36%, while labor productivity for the economy as a whole went up 25%.

But the pendulum has not swung totally in the direction of material incentives; mass mobilization and voluntarism are still features of Cuban life. The work force continued to be able to respond to national and local challenges — the rapid recovery from the effects of the rust disease outbreak and the successful campaign to reduce the consumption of imported energy stand out. There may be limits to this, however. With factory workers increasingly exhibiting a high level of training and skill, there may be resistance to the call for them also to work in the fields. Such situations exist now, for example, when, after the completion of the harvest, additional hands are needed to replant and weed the canefields (*Granma*, 6/2/86, p. 1). The factory workers are mobilized, but at what economic and social cost?

(d) *The liberation of women*

Recent developments in the Cuban sugar industry also help to clarify issues surrounding the role of women in postrevolutionary Cuban society. Some observers of the Cuban scene emphasize the continued existence of a "second shift" for women, the low levels of employment of women in certain areas, the low labor force participation rates of women compared to those of men, the unequal political role of women and the continued expression of chauvinist behaviors in the culture, and criticize the efforts of the Revolution in this area. Others emphasize the gains that women have made relative to their position in prerevolutionary society, or in other systems. Deere, for example, argues that the Cuban case is an exception to the general rule in Latin America, according to which agrarian reform directly benefits only the men (Deere, 1984b).

The recent experience of the sugar industry contains evidence for both sides. The data described above show that complete equality has not been attained, yet the achievements have been immense. Before the Revolution, the sugar sector was one in which women were severely hampered. Today women are playing central roles, especially in the cooperatives and in technical positions within the factories. In looking at the data, and comparing it to both the prevolutionary situation and the situation in other sugar producing countries, the advances by women are very impressive.

(e) *Decentralization*

Along with the growing utilization of material incentives, Cuban officials have attempted to promote a more decentralized model of economic organization. They have also attempted to maximize organizational coherence and efficiency at the local level (Zimbalist, 1985, p. 223). In the sugar sector during the first half of the 1980s, this was reflected in the establishment of the agro-industrial complexes, which rationalized and integrated production functions for the first time; in the formation of permanent production brigades, which appeared to extend the decentralization principle even further to groups of workers; and to self provisioning which stimulated enterprises to greater levels of self-sufficiency. Whether the reduction in growth rates between 1981 and 1985 is related to teething problems experienced in these areas is unclear. The government, however, seems generally satisfied and is continuing to support the complexes, the brigades and self provisioning.

NOTES

1. For comparative purposes, the percentage of agricultural land in state farms in the USSR is 65%, in Bulgaria 16%, in Czechoslovakia 30%, Rumania 30%, and East Germany 7% (all 1975 figures). The figures for socialist Third World countries are even lower. For example, state farms comprise only 3% of the land in agricultural use in the centrally-planned countries of Asia (China, Mongolia, Korea, Vietnam and Laos) (Galeski, 1985, p. 20). Deere reports that 27% of the farmland in Algeria, 24% in Nicaragua, 12% in Angola, and 10% in Yemen is in the state sector of those countries (Deere, 1984a, p. 22 and Appendix B; all 1980 figures except Nicaragua which is for 1983). An important caveat should be noted here. That is that most of the centrally-planned economies also have fairly substantial collective farm sectors and in these countries there are often important similarities in the management of the collective farms and the management of the state farms. Some argue that it is really just one large state sector.

2. During this period, most of the campesinos were organized into peasant associations (National Organization of Small Peasants — ANAP); a smaller percentage (less than one-third of all peasant households) also belonged to credit and service cooperatives. Only 3% belonged to producer cooperatives (Deere, 1984a, p. 55).

3. The government established a number of incentives to motivate the campesinos to agree to pool their resources. CPA members were eligible to enter the social security system, received preferential access to equipment, inputs and credit (compared to independent campesinos); and received preferential access to scarce construction materials with which to build collective and individual facilities, such as housing, schools, and day care centers (Deere, 1984a, p. 57).

4. A closer look at the data (Table 2) shows that between about 1983 and 1986, the number of cooperatives and cooperative members *declined*, while the average farm size increased. This followed a period of very rapid growth in the amount of land integrated into the cooperative sector. This trend toward amalgamation may have been checked, given Fidel's comment at the Third Party Congress: "The occasional tendency to establish very large cooperatives should be avoided considering the lack of managerial and technical cadre among the farmers" (Castro, 1986, pp. 11–12).

5. Pérez (1986) reports 427 cane cooperatives, Leiva (1985), 428 and *Granma* (10/12/85), 432. These figures are a sharp drop from the 493 cane cooperatives reported by Benjamin (1984) for 1983. With the cane cooperatives as with the others, there was a strong tendency toward increasing size and reduced numbers after about 1983. This coincided with a period of serious production problems in the campesino cane sector (see section 2.b). Note that not all of the land held by the cane cooperatives was allocated to the production of sugar cane (Batista, 1985, p. 6).

6. Such foreign matter taken in to the factories with the cane harms equipment, falsely adds to sugar-cane production statistics, and takes up valuable space in the vehicles that transport the cane.

7. Cuba has as many cane combines in operation as all the other sugar-producing countries combined (*Granma Weekly Review*, 6/8/86, p. 3).

8. For more details on these points and a full discussion of harvest mechanization in Cuba, see Edquist (1985).

9. The margin averaged 8% between 1971 and 1975.

10. Sugar cane is a perennial which can be reaped for 10 or more years without replanting. Yields, however, decline with age. Replanting, on the other hand, is labor intensive and expensive.

11. This commitment must be put in context. While sugar industry investment between 1976 and 1980 was more than twice the figure for 1966–70, total investment in Cuba was *three times* the figure for the earlier period. Sugar industry investment thus declined as a percentage of total investment. I have not seen comparable figures for the more recent period but I suspect they were similar (Castro, 1981, p. 16).

12. Despite the construction of new factories, the modernization of others, and a general improvement in infrastructure in recent years, the factories still require much work. Still in use, for example, are 350 mills dating from the 19th century (*Mundo Azucarero*, September 1985, p. 2).

13. This conclusion is based on a sampling of all *Granma* issues between January 1985 and February 1986.

14. There are about 60,000 women in the union of sugar workers, of which 15,000 work in the factories. Few women cut the cane or operate field machinery; hence most of the rest are field workers (*Granma*, 3/14/85; Pérez, 1986, p. 2).

15. The distinction between professional cane cutters, those whose full-time occupation is in the sugar industry, and volunteer cutters who come in only for the harvest, is not clearly drawn. For one thing many of the volunteers work regularly as cane cutters during the harvest, and for successive harvests. For another, their productivity is often higher than the professionals'. In the 1982 *zafra*, for example, only 26% of the Millionaire Brigades were made up of full-time sugar workers (*Latin America Commodities Report*, 7/16/82, p. 3).

16. Are the membership figures representative of all 153 agro-industrial complexes? The data are obviously unsystematic. There is no reason, however, to think that these three complexes were singled out because of unusually high Party membership levels. The stories in which the data were reported discussed sugar operations, not Party affairs. The Communist youth membership figures are reported for two of the three complexes.

17. Cane cutters, for example, were eligible for a 10% premium for fulfilling their work norm — 2.99 tons per day over a two week period — and for 80% attendance (*Latin America Commodities Report*, 9/26/80, p. 5; the norm had been 2.76 tons per day before the wage reform).

18. In 1982, for example, 1,014 tourist trips to CMEA countries at half price were made available to the workers as incentives (*Latin America Commodities Report*, 11/12/82, pp. 6–7).

19. *Anuario* (1983, pp. 109, 141) reports a narrower gap between sugar and non-sugar workers in industry, with some narrowing of the gap in 1983.

20. Deere argues that none of the 10 countries in the Third World which have established state farms as part of their transition to socialism, including Cuba, has "had a successful experience in participatory management" (Deere, 1984a, p. 35).

21. Of course, there is considerable conjecture that these calamities were anything but natural. Some see the hand of the Central Intelligence Agency, for example, in the various mysterious outbreaks of strange diseases in the late 1970s and early 1980s, including the rust disease attack on the Barbados 4362 variety of sugar cane.

REFERENCES

Batista Almaguer, Cornelio, "Eficientes Cooperativistas Cañeros," *ANAP*, Vol. 25, No. 10 (October 1985).

Batista Almaguer, Cornelio, et al., "Encuentros Cooperativos Por Ramas: De Caña en Matanzas," *ANAP*, Vol. 25, No. 7 (July 1985).

Benjamin, Medea, Joseph Collins, and Michael Scott, *No Free Lunch: Food and Revolution in Cuba Today* (San Francisco: Institute for Food and Development Policy, 1984).

Castanãeda, Mireya, "Sugarcane: Using every little bit," *Granma Weekly Review*, Vol. 21, No. 5 (2 February 1986).

Castro, Fidel Ruz, "Main report to the 2nd Congress of the Communist Party of Cuba," *2nd Congress of the Communist Party of Cuba: Documents and Speeches* (Havana: Political Publishers, 1981).

Castro, Fidel Ruz, "Main report to the 3rd Congress of the Communist Party of Cuba," *Granma Weekly Review*, Vol. 21, No. 7 (16 February 1986).

Comité Estatal de Estadísticas, *Anuario Estadístico de Cuba* (Havana: various years).

Comité Estatal de Estadisticas, *Estadísticas Quinquenales de Cuba, 1965–1980* (Havana: 1982).

"Cuba: Una industria azucarera nacional," *Mundo Azucarero*, Vol. 8, No. 3 (September 1985).

Deere, Carmen Diana, "Agrarian reform and the peasantry in the transition to socialism in the Third World," Working Paper No. 31, The Helen Kellogg Institute for International Studies (Notre Dame: University of Notre Dame, December 1984a).

Deere, Carmen Diana, "Rural women and state policy: The Latin American agrarian reform experience," *World Development*, Vol. 13, No. 9 (1984b).

Edquist, Charles, *Capitalism, Socialism and Technology* (London: Zed Books, 1985).

Food and Agriculture Organization, *FAO Monthly Bulletin of Statistics* (Rome: FAO, various issues).

FAO, *FAO Production Yearbook* (Rome: FAO, various issues).

Fuller, Elaine, "Changes expected in Cuba's economy," *Cubatimes* (January/February 1985).

Galeski, Boguslaw, "Land and manpower in collective farming in the world," Typescript (n.d., circa 1985).

Gómez, Orlando, "Sus Cañas Son Tambien Vitales," *Granma* (23 October 1985).

International Sugar Organization, *Sugar Year Book* (various years).

Latin America Commodities Report (London: Latin America Newsletters Ltd., various issues).

Leiva, Chongo, "Los nuevos precios de la caña," *ANAP*, No. 10 (1985).

López Moreno, José A., "Report on fulfillment of the 1985 Plan for Economic and Social Development," Report to the National Assembly, *Granma Weekly Review* (12 January 1986).

Machado, Ricardo J., "Los Campesinos de Madruga," *ANAP*, Vol. 26, No. 1 (January 1986).

"Mas Caña Con Los Nuevos Precios," *ANAP*, Vol. 26, No. 1 (January 1986).

National Foreign Assessment Center, *The Cuban Economy: A Statistical Review* (Washington, DC: Central Intelligence Agency, 1981).

Pérez Rojas, Niurka, "Women in the Cuban sugar agro-industry," *Granma Weekly Review* (12 January 1986).

Pollitt, Brian H., *Revolution and the Mode of Production in the Sugar-Cane Sector of the Cuban Economy, 1959–1980 — Some Preliminary Findings*, Occasional Papers No. 35, Institute of Latin American Studies (Glasgow: Glasgow University, 1981).

Ramírez Cruz, José, "Desarrollo Economico-Social Del Campesinado," *ANAP*, Vol. 25, No. 8 (August 1985).

Varela Pérez, Juan, "66 Percent Mechanization of Canecutting For 1985–86 Harvest," *Granma Weekly Review* (27 October 1985).

Zimbalist, Andrew, "Cuban economic planning: Organization and performance," Sandor Halebsky and John Kirk (Eds.), *Cuba's Twenty-Five Years of Revolution, 1959–1984* (New York: Praeger, 1985).

·5·

Cuban Industrial Growth, 1965-84

ANDREW ZIMBALIST

1. INTRODUCTION

From 1980 to 1985 real per capita GDP in Latin America fell at an average annual rate of 1.7% (19 countries excluding Cuba), according to an ECLA calculation based on official government statistics.[1] In sharp contrast, again according to official statistics, constant price per capita GSP in Cuba grew at an average annual rate of 6.7% during the quinquennium.[2] If the official Cuban statistics could be accepted at face value, they would *prima facie* represent a very impressive economic performance.

However, methodological differences, administered prices, procedural irregularities and opaqueness, double counting, new product pricing practices, among other things, have engendered caution and suspicion among Western specialists when interpreting official national income figures from centrally-planned economies. Many comparative economists have employed techniques such as Bergson's adjusted factor cost pricing to make the national income estimates of the MPS methods compatible with those of the Western SNA methods.[3] To be sure, the recently published International Comparisons Project of the World Bank made such computations for eight CMEA countries.[4]

As part of this project, two economists, Carmelo Mesa-Lago and Jorge Pérez-López, were asked to analyze Cuban economic performance. In their report, however, not only did they cite a lack of sufficient data to generate their own estimates of Cuban national income and its growth but they cast a broad doubt on existing independent estimates of Cuban economic performance.[5] Whereas their criticisms of these studies were on the whole well taken, their own discussion of Cuban national income methodology and statistics was amply flawed.[6]

Subsequent to his participation in the World Bank project, Pérez-López was hired by Wharton Econometric Forecasting Associates (WEFA), itself under contract to the US State Department, to develop estimates of Cuban economic growth. This WEFA study was completed in November 1983 and then updated in a 1985 paper by Pérez-López.[7] Both the initial WEFA study and its update present a vigorous challenge to the rosy figures of the Cuban government. The WEFA estimates have, in turn, been published by the CIA in its most recent *The Cuban Economy: A Statistical Review*[8] and have been cited extensively in US press coverage of the Cuban economy. Notably, in the September/October 1985 issue of *Problems of Communism* the well-known Cubanologist Jorge Domínguez wrote: "Cuba's statistical system has yet to generate credible data about the obviously economically troubled 1980–82 period, for which official figures unconvincingly suggest an economic boom."[9]

These conflicting positions provoked a debate published in three 1985 issues of *Comparative Economic Studies* between Mesa-Lago and Pérez-López, on the one hand, and Swedish economist Claes Brundenius and myself, on the other. In this paper, after briefly summarizing the existing studies, I shall present and evaluate new estimates of Cuban industrial growth. These estimates are based on more comprehensive price data and more meaningful branch weights

---

The author gratefully acknowledges the generous support of the Jean and Harvey Picker Fund.

than those used in the WEFA study. They tend to lend credibility to the official record.

2. THE WEFA STUDY

In January 1981 a major wholesale price reform was put into effect in Cuba. According to data derived from the 1984 *Cuban Statistical Yearbook*, the implicit rate of wholesale price inflation in 1981 was 9.9%.[10] WEFA argues, however, that the official figures which report a real GSP increase of 16.0% in 1981 do not fully account for the price increases occasioned by the reform. Thus, the use of constant prices would provide a truer measure of the increase in real output in 1981.

Due to the absence of sufficient Cuban price data and questions about the meaningfulness of Cuba's administered prices, WEFA employs a set of Guatemalan prices from 1973 and supplements this set with US prices. Since Guatemalan industry at the time enjoyed average tariff protection in excess of 25% as a member of the Central American Common Market and, affecting non-tradables, Guatemala's factor endowments were markedly different from Cuba's, some skepticism about the appropriateness of this price set is warranted. Acknowledging the difficulties of data collection, however, let us look further into the methods and results of the WEFA study.

WEFA used the so-called bottom-up method and applied its Guatemalan/US price set to the 206 industrial products whose physical output is reported in the *Cuban Statistical Yearbook* for each of Cuba's 21 industrial branches.[11] More precisely, Laspeyres quantity indexes were calculated as follows:

$$I_{b_i} = \frac{\sum_{j=1}^{n} (p_{j0})(q_{jt})}{\sum_{j=1}^{n} (p_{j0})(q_{j0})} \qquad (1)$$

where I_{b_i} is the output index for branch b_i, n is the number of products in WEFA's sample for the branch, p_{j0} is the base period price for the jth product in branch b_i, q_{jt} is the physical output of product j in the year t (where t runs from 1965 through 1980) and q_{j0} is the physical output of product j in the base year. Thus, activity indexes were estimated for each branch dating back to 1965.

The branch indexes were then aggregated by using hybrid estimates for branch value added. The majority of these estimates were generated roughly as follows: ECLA estimates for branch value added (VA) in 1963 (based on JUCEPLAN figures for branch gross value added in 1961) were divided by estimated branch gross value of output (GVO) for that year; assuming the branch VA/GVO ratios to be constant between 1963 and 1975, the 1963 ratio was multiplied by branch GVO in 1975 yielding an estimate for value added in that year.[12] Specifically,

$$I = \sum_{j=1}^{21} (I_{b_i})(W_{b_i}) / \sum_{j=1}^{21} W_{b_i} \qquad (2)$$

where I is the activity index for industry and W_{b_i} is the value added estimate for the ith branch.

There are significant problems both with WEFA's intra- and inter-branch weighting procedures. WEFA's price set covers only 128 out of the 206 products for which they have yearly physical output data, or 62.1% of their product sample. Since the Guatemalan economy is significantly less industrialized then the Cuban economy, the 1973 Guatemalan price sample seems systematically to leave out many of Cuba's newest and most dynamic industrial product groups. This observation also holds for the 206 products included in Cuba's statistical yearbook as this list has only incorporated eight new products since 1970 and none since 1978. This double sampling bias produces a downward skewing in the WEFA estimates of branch activity indexes.

In Table 1 I reproduce an updated comparison of the WEFA output index for five branches with an index generated by Brundenius and myself based on 1981 Cuban wholesale prices.[13] In four of the five cases the increased product coverage allowed by the Cuban price set resulted in significantly higher estimates for branch output growth. In the fifth case, the construction materials branch, the expanded coverage estimate for post-1974 growth is markedly higher. The questions remain, of course, as to whether (a) this higher growth is due to product coverage or to a growth bias in the Cuban price set and (b) a similar tendency prevails in the other industrial branches. I shall return to these questions below.[14]

WEFA's inter-branch weights also present a series of problems ranging from the assumption of constant VA/GVO over 1963–73 to their hybrid fabrication. The most significant source of bias in these weights, however, appears to be in

Table 1. *Comparison of estimated growth indexes for selected branches* (1974 = 100)

	1965	1970	1974	1980	1984
Construction materials (12 products in sample)					
WEFA (8 of 12 products)	22	27	100	111	—
Brund/Zim. (12 of 12)	36	37	100	131	150
Metal products (8 products in sample)					
WEFA (2 of 8 products)	16	31	100	46	—
Brund/Zim. (8 of 8)	21.8	38	100	184	283
Non-electrical machinery (11 products in sample)					
WEFA (5 of 11 products)	10	21	100	74	—
Brund/Zim. (9 of 11)	2	21	100	205	258
Electrical machinery (8 products in sample)					
WEFA (6 of 8 products)	58	34	100	129	—
Brund/Zim. (8 of 8)	60	45	100	162	216
Chemicals (29 products in sample)					
WEFA (11 of 29 products)	57	77	100	92	—
Brund/Zim. (23 of 29)	50	64	100	188	197

the use of weights based on *precios del productor* (inclusive of turnover taxes) for certain branches instead of on *precios de empresa* (exclusive of turnover taxes). A stunning example of the resulting distortion is the 26% weight assigned to the Beverages and Tobacco branch. This weight, over a quarter of all industrial output in Cuba, is approximately three times the weight assigned to the food industry, three times the weight given to chemicals and four times the weight given to the sugar products branch.

As can be readily appreciated in Table 2, 76% of all industrial turnover taxes fall on the Beverages and Tobacco branch. In fact, the 1981 net value added weight exclusive of turnover taxes for this branch is 4.5%, i.e., the WEFA weight is 5.8 times too high. Since according to WEFA's limited product coverage the output in this branch decreased between 1965 and 1980, this weighting distortion produces a powerful downward bias in the WEFA estimate of industrial growth.

3. NEW ESTIMATE

Brundenius and I obtained from the Cuban State Statistical Committee (a) a set of official 1981 wholesale prices and (b) official branch net value added figures at both enterprise and producer[15] prices. Applying the bottom-up method to this data we arrived at the new estimates for industrial growth reported in Method A of Table 3. Since we received no price data for six of the 21 industrial branches, the index and annual growth rate estimated in Method A is based on 15 branches which accounted for 77.4% of net industrial value added in 1981.

The most significant omission is that of the sugar products branch with a 1981 net value added weight of 10.2%. Using the official current price GVO series and the official physical output series from the 1984 statistical yearbook, I was able to estimate that average 1981 wholesale price increases in this branch equalled 76%. Deflating the GVO series by this figure I arrived at an estimated constant price GVO series for the sugar products branch. Following a similar procedure I estimated a constant price GVO series for the fuel branch as well. The "adjusted growth rate" of Method A incorporates these two branches to the original 15. It, thus, represents 17

Table 2. *Turnover taxes as a share of gross output in industry*, 1981 (million pesos)

Branch	Turnover taxes*	Gross value of output at enterprise prices	Share of turnover taxes in GVO	Share of turnover taxes in GVO in USSR, 1972†
Electrical energy	0	452.2	0	4.1%
Fuel	0	512.6	0	15.1%
Mining & ferrous metallurgy	0	108.5	0	0.2%‡
Mining & non-ferrous metallurgy	0.3	135.7	0.2%	
Non-electrical machinery	0.2	553.4	0.04%	3.7%
Electric machinery	0	100.6	0	
Metal products	0	141.4	0	0.2%‡
Chemicals	196.8	427.0	46.1%	3.7%
Paper products	0	136.0	0	0.9%
Graphics	0	81.1	0	
Wood products	2.0	126.6	1.6%	
Construction materials	0	369.1	0	1.4%
Glass & ceramics	0.1	32.2	0.3%	18.0%
Textiles	0	163.4	0	
Apparel	0.8	195.2	0.4%	
Leather	0.4	152.9	0.3%	
Sugar	9.0	1,421.5	0.6%	20.8%
Food	213.8	1,653.8	12.9%	
Fish	0	211.8	0	
Drink & tobacco	1,359.6	350.3	388.1%	
Other	6.1	403.2	1.5%	7.6%

*Calculated by the author as the difference between official net value added in *precios del productor* and *precios de empresa*.
†From Pitzer (1982), p. 35. Although not specified, I assume Pitzer's measure of gross output is in enterprise prices, i.e., exclusive of turnover tax.
‡The average turnover tax of these three branches is 0.2%.

of 21 branches accounting for 90.4% of total net industrial value added in 1981 and stands as my best estimate in this study. (Due to greater product heterogeneity and the change in Cuba's classification scheme in the mid-1970s, I was unable to make reliable constant price GVO estimates for the remaining four branches.)

As shown in Table 3, my adjusted growth rate estimate lies in between the WEFA estimate for the period 1965–80 (the years covered in the initial WEFA study) and the official figure. Nevertheless, my estimate (6.5%) comes much closer to the official figure (7.1%) than to WEFA's (2.3%). Futhermore, my estimate for growth in 1981 (11.7%) corroborates the impression of a growth spurt given by the official figure for real industrial growth, as measured by gross value of output, in that year (18.0%). In the WEFA update Pérez-López actually estimated that industrial growth was negative in 1981 (−1.8%) suggesting that the official price deflator grossly underestimated actual inflation. Whereas the relatively modest difference between my constant price estimate for 1981 growth and the official figure for the same year might be attributable in part to an underestimated official price deflator, I believe it is more probably a function of the factors to be discussed in the next section.[16]

Methods B and C use the same set of 1981 Cuban prices as Method A but they employ different inter-branch weights. The rather substantial differential in their estimated growth rates from that in Method A suggests the sensitivity of these estimation procedures to the underlying weighting scheme. Recalling the greater importance of alternate weighting schemes for Soviet growth estimates during its period of fast industrialization (pre-1950)[17] and observing the rapid shift in Cuba's industrial product mix since 1970 at least, this sensitivity is hardly surprising.

Table 3. *Cuban industrial growth* (constant prices)

	1965	1970	1974	1980	1981	1982	1983	1984	1965–80	1980–84	1965–84
Method A*											
Index (1974 = 100)	52.5	65.3	100	149.4	165.2	162.1	171.8	182.3			
Annual growth rate	—	—	—	—	10.6	−1.9	6.0	6.1	7.2	5.1	6.8
Adjusted growth rate†	—	—	—	—	11.7	—	—	—	6.5	5.5	6.3
WEFA (Pérez-López)											
Index	81	82	100	114	112	119	—	—	—	—	—
Annual growth rate	—	—	—	—	−1.8	6.3	—	—	2.3	—	—
Official (GVO)											
Index	54	74.2	100	150.5	177.6	185.9	194.4	211.6	—	—	—
Annual growth rate	—	—	—	—	18.0	4.7	4.6	8.8	7.1	8.9	7.5
Method B‡											
Index	76.1	85.3	100	133.7	142.6	138.5	147.0	159.4	—	—	—
Annual growth rate	—	—	—	—	6.7	−2.9	6.1	8.4	—	4.5	4.0
Method C§											
Index	75.9	88.2	100	130.5	137.9	139.7	146.4	153.0	—	—	—
Annual growth rate	—	—	—	—	5.7	1.3	4.8	4.5	—	4.1	3.8

*1981 Cuban wholesale prices for product weights; 1981 net value added at enterprise prices (turnover tax not included) for branch weights. Fifteen of 21 branches included, accounting for 77.4% of net industrial value added.
†Adds GVO estimates at constant prices for two additional branches: fuel (weight = 0.028); sugar products (weight = 0.102). Seventeen branches included, accounting for 90.4% of net industrial value added.
‡1981 Cuban wholesale prices for product weights; branch weights from WEFA.
§1981 Cuban wholesale prices for product weights; sample GVO in each branch summed.

In the final section of this paper I shall attempt to quantify the sources of the differentials in the various growth estimates under study. In the next section I shall turn to consider the possible forces imparting an upward or downward bias to my estimate.

4. EVALUATION OF ESTIMATE

(a) *Sources of upward bias*

There are three reasons why one might suspect my estimate to be biased upward. The first is the omission of branches with slower growth. The five candidates here are fuel, sugar products, graphics, mining and non-ferrous metallurgy and apparel. The first two have already been included in my adjusted estimate. The non-inclusion of graphics does not appear to be a serious problem. On the one hand, this branch had a net value added weight of only 1.2% in 1981. On the other hand, the slow growth of graphics production since 1976 suggested by figures in the statistical yearbook is misleading as it only includes output by economic units in the material production sphere. Hence, official government publications, journals of academic institutions, magazines and newspapers issued by public organizations, all of which have experienced an obvious boom over the last 10 years, are excluded.[18] Furthermore, applying a 1967 set of prices from Peru with coverage of four out of five products, graphics output grows at an annual rate of 10% over 1965–80.[19] This, of course, suggests that excluding graphics might create, if anything, a downward bias to my estimate.

Mining and non-ferrous metallurgy had a net value added weight of 2.3% in 1981 and likely grew at a pace below the overall industrial average. According to the WEFA Guatemalan price set, the annual 1965–80 growth rate of this branch was 1.3%. Based on the 1967 Peruvian prices, however, the rate would be 6.1%.

The 1981 value added weight of the apparel branch was 3.2%. This branch according to the WEFA price weights grew at an average annual rate of 2.6%, well below the industrial average. Counterbalancing the effects of these latter two branches, however, is the fish products branch (1981 VA weight of 2.9%), which was also excluded from my estimate due to a lack of Cuban price data. In WEFA's price sample only three of the seven products of this branch were covered and the estimated branch growth rate was 7.3%. Applying ton weights to all seven products (i.e., all fish products are given an equal price per pound), the growth rate of this branch is 8.9%. On balance, it is difficult to conclude that the excluded branches impart a systematic bias to my estimate, especially given their total VA weight is below 10% and given the incomplete and questionable price data underlying any effort to measure their growth.

The second reason is the possibility of distortions arising from new product pricing.[20] It appears to be a standard practice in the Soviet Union to price new products (which are sometimes new in only a cosmetic sense) high enough to fully and immediately recover all research and development costs. Further, it has been claimed that Soviet pricing authorities are too overwhelmed with the enormity of their task to pay close attention to most new products and, hence, simply ratify the prices requested by the producing enterprises. In theory this new price should be lowered once the initial R&D costs are recovered. In practice, the price is generally maintained. The resulting distortion to growth rates is twofold. First, price increases for products tend to exceed quality increases and this biases growth estimates based on Soviet prices upward. Second, new products which tend also to be more dynamic and faster growing receive higher price weights, also biasing the growth rate upward.

Neither pricing regulation documents, nor published articles, nor discussions with Cuban pricing authorities have revealed any theoretical inclination to price new produts in this way. Cuban price officials clearly feel overwhelmed and they too receive price proposals and exhortations from enterprises, but I have no evidence that they simply ratify enterprise requests for higher prices. Should such practices be in effect they could impact on my measure in three ways: (a) in the measurement of constant price GVO for the fuel or sugar products branch; (b) in the pricing of individual products or, (c) indirectly, in the branch value added weights which are based on Cuban prices. Since I am using constant year prices, however, there is no danger that the same product (only cosmetically altered) is assigned a higher price in the middle of the series. Again, I have no evidence to suggest a distortion of this type, but it is nevertheless appropriate to acknowledge this potential problem.

The third possible source of upward bias resides in my use of late year branch weights. As we shall explain below, late year *price* weights for individual products (intra-branch weights) would tend to impart a downward growth bias if Cuban prices behave conventionally. The employed value added branch weights, however, are a multiplicative function of both 1981 prices and 1981 quantities, tending to impart a higher weight for any given price set to branches with more rapid growth than would the use of branch value added weights from an earlier year. In the capital goods related branches with negligible output in 1965 and average annual growth rates in excess of 15% since 1970, the potential bias here is significant.

(b) *Sources of downward bias*

There are several reasons to suspect a downward bias in my growth estimate. First, in using 1981 prices and value added weights I am applying late year weights. If Cuban prices behave in conventional ways, then according to the standard index number formulation the use of prices from the end of the period would bias the resulting growth estimate in a downward direction. On the other hand, the often cited bias of the bottom-up procedure in underestimating quality improvements would be minimized by the use of late year prices if Cuban prices are adjusted to reflect quality gains over time.

Second, there appears to be little question that the products reported in the Cuban *Statistical Yearbook* and which form the basis for both my estimate and the WEFA estimate seriously underrepresent new products. As mentioned above, only eight new products have been added to the list since 1970 and none since 1978. New products and product groups tend to grow more rapidly and their exclusion from the sample, therefore, engenders a downward bias. Brundenius and I were able to explore this impact in detail for three branches on the basis of additional output and price information supplied by the State Statistical Committee to Brundenius for a study of capital goods output.[21] In Table 4 four different estimates of capital goods production in the metal products, electrical machinery and non-electrical machinery branches are reported. The WEFA estimate is based on a total of 13 goods in these three branches and 1973 Guatemalan prices. The Brundenius/Zimbalist (1) and (2) estimates use 1981 Cuban prices and are based on 25 products (appearing in the *Statistical Yearbook*) and over two hundred products, respectively. The official Cuban estimate is of course based on all products in these branches and, reportedly, 1981 prices. The underlying tendency is clear: the more products included in the estimate, the higher the growth rate. The

Table 4. *Comparison of estimated growth indexes for engineering goods branches** (1980 = 100)

	1980	1981	1982	1983
WEFA (projected using WEFA methodology)	100	123	86	91
Brundenius/Zimbalist (1)	100	120	111	127
Brundenius/Zimbalist (2)	100	122	138	163
Official Cuban	100	124	139	164

*These data refer to three branches: metal products, electrical machinery and non-electrical machinery. The B/Z(2) estimate includes only the capital goods (excluding all consumer durables) in these three branches. In value terms it accounts for approximately 37% of the production in these three branches according to the official statistics.

contrast between the second and third estimates is sharp and is based on the inclusion of products not listed in the *Statistical Yearbook*.

Two additional inferences are in order: the Cuban State Statistical Committee does not select products for *Yearbook* listing in order to give the impression of higher economic growth and the similarity of Brudenius/Zimbalist (2), measured in constant 1981 prices, and the official Cuban estimates suggests that there is no hidden inflation buttressing the official figures for these branches.

A third reason my estimate might be downward biased lies in the nature of Cuban value added weights. Cuban capital goods prices are notoriously low. Since capital goods output grew at 16.4% in real terms between 1965 and 1983,[22] the low relative prices denote low value added weights for the three related branches and, hence, would create a downward bias in the overall industrial growth rate.

There is also reason to suspect that the use of net value added weights (net of capital consumption allowance) instead of gross value added might impart a downward bias if the faster growing branches also tended to be more capital intensive. As is common in the MPS system, value added is defined in net terms and this was the only data available to us. Efforts to estimate depreciation from branch capital stock statistics were frustrated by irregular and, I believe, unreliable data. Running my estimates for branch capital/output ratios on branch growth rate yielded insignificant correlation coefficients. If better capital stock series data become available, it would be interesting to explore this relationship further.

A fourth possible source of downward bias is that the physical output series employed represents gross (double counted) output. According to official Cuban statistics, Net Material Product or National Income has grown more rapidly than Gross Social Product (by 0.6 percentage points per year for the whole economy and by 1.6 percentage points per year for the industrial sector) over the last 10 years. Since the same, constant prices are used in each series, this implies that gross output has grown more slowly than net output; put differently, the use of intermediate inputs per unit of output has declined. The use of gross physical output, then, would create a downward bias as well.

The final possible source of downward bias resides in my use of constant price GVO estimates for the fuel and sugar products branches. The Cubans switched their valuation methodology for measuring gross value of output from gross turnover (*circulación completa*) to enterprise exit (*a la salida de empresa*) in 1976. Enterprise exit, of course, eliminates some double counting for inputs produced in the same enterprise and, thus, results in lower output valuation. The Cubans evaluated 1974 GSP using both methodologies. With enterprise exit GSP was 7.6% lower than with gross turnover. Since the GVO for both fuel and sugar in 1965 were evaluated by the gross turnover method and in 1980 by the enterprise exit method, this would bias the base year figure upward and the resulting growth estimate downward. The distortion here, however, as in the fourth source, is likely to be slight. In any event, it appears evident considering all these factors that the potential for downward bias in my estimate is equal to or stronger than the potential for upward bias.

In Table 5 above branch weights net of turnover tax in both GVO and VA terms are presented. Column four records the extent to which a branch's value added weight exceeds its gross output weight. The simple correlation coefficient between columns three (branch growth rates) and column four is 0.47, significant at the 0.06 level and denoting that the use of net value added weights results in higher growth estimates than gross output weights.

5. SOURCES OF DIFFERENTIALS IN GROWTH ESTIMATES

In this section we follow the lead of Stanley Cohn[23] and endeavor to quantify the impact of (a) Cuban prices, (b) product coverage and (c) branch weights on the various estimates for Cuban industrial growth. In Table 6 four separ-

Table 5. *1981 branch shares and growth rates*

	(1) Branch weight, percent of GVO	(2) Branch weight, percent of value added	(3) Average annual growth rate, 1965–84 (constant prices)*	(4) (2) − (1)
1. Electrical energy	5.9	12.2	7.9	6.3
2. Fuel	6.3	2.8	(3.3)	−3.5
3. Mining & ferrous metallurgy	1.4	1.5	14.7	0.1
4. Mining & non-ferrous metallurgy	1.8	2.3	—	0.5
5. Non-electrical machinery	7.2	14.8	32.1	7.6
6. Electrical machinery	1.3	2.3	7.0	1.0
7. Metal products	1.8	3.1	14.4	1.3
8. Chemicals	5.5	4.3	7.5	−1.2
9. Paper products	1.8	2.8	5.9	1.0
10. Graphics	1.0	1.2	—	0.2
11. Wood products	1.6	3.7	−1.3	2.1
12. Construction materials	4.8	6.0	7.8	1.2
13. Glass & ceramics	0.4	0.4	6.7	0
14. Textiles	2.1	2.0	2.6	−0.1
15. Apparel	2.5	3.2	—	0.7
16. Leather	2.0	4.3	0.4	2.3
17. Sugar	18.4	10.2	(3.6)	−8.2
18. Food	21.4	10.2	2.8	−11.2
19. Fish	2.7	2.9	—	0.2
20. Drink & tobacco	4.5	4.5	0.4	0
21. Other	5.2	5.4	5.7	0.2

*Parentheses indicate a different estimation procedure, described in text. They do not denote negative rates.

Table 6. *Impact of Cuban prices, product coverage, and branch weights on industrial growth rates*

	1965	1966	1969	1970	1971	1974	1975	1976	1980
Estimate A: 1981 Cuban prices, WEFA branch weights, WEFA product coverage	77	85	99	87	79	100	110	106	114
Estimate B: 1973 Guatemalan prices, WEFA branch weights, WEFA product coverage	81	83	91	82	82	100	109	106	113
Estimate C: 1981 Cuban prices, WEFA branch weights, extended coverage	76	83	97	85	78	100	108	108	134
Estimate D (method A): 1981 Cuban prices, Cuban net value added branch weights, extended coverage	52.5	58.9	63.5	65.3	77.5	100	112.3	112.2	149.4

ate estimates are presented. Estimate A is based on 1981 Cuban prices, WEFA branch weights and WEFA's restricted product coverage. Estimate B differs in method from Estimate A only in its use of Guatemalan, instead of Cuban, prices. Esimate C departs from Estimate A only in its extended product coverage, corresponding to the greater coverage of the Cuban price set. Finally, Estimate D differs from Estimate C in its use of Cuban net value added weights from 1981.

Table 7 depicts the average annual growth rates of each estimate. By comparing the in-

Table 7.

	Average annual growth rate 1965/66–1980
Estimate A	2.3
Estimate B	2.2
Estimate C	3.5
Estimate D	6.8

cremental change with the difference between the highest and lowest estimates Table 8 shows the contribution of the various factors to explaining the variation in growth rates. The use of Cuban rather than Guatemalan prices has virtually no impact. Indeed, if it were possible to adjust for the fact that the Cuban prices are from 1981 and the Guatemalan price are from 1973, one might find that if anything the Cuban prices produce a downward growth bias. This would be consistent with the previous discussion of low

Table 8.

	Percent difference between low and high estimate accounted for by
Cuban prices	2.2*
Product coverage	26.1†
Branch weights	71.7‡

*Equals (Est. A − Est. B) ÷ (Est. D − Est. B).
†Equals (Est. C − Est. A) ÷ (Est. D − Est. B).
‡Equals (Est. D − Est. C) ÷ (Est. D − Est. B).

prices for capital goods. Extended product coverage accounts for 26.1% of the growth estimate differential and the choice of branch weights accounts for nearly three-quarters of the differential. Again, this latter finding is to be expected given the rapid and substantial transformation of Cuba's product mix since 1965.

6. CONCLUSION

This paper has employed the bottom-up method with Cuban late year prices and value added weights to generate a new, independent estimate of Cuban industrial growth from 1965 to 1984. The price set enabled coverage of branches accounting for 90.4% percent of industrial value added, once the constant price GVO was estimated for two additional branches. On the whole, the estimates of this study are close to, albeit below, the official growth figures. In my judgment, this result lends heightened credibility to the implicit price deflators used in the official Cuban statistics as well as to the overall presentation of economic data in the Cuban *Statistical Yearbook*.

Due to a lack of data on industrial subsidies and no access to an input–output table, I was unable to develop adjusted factor cost prices. Adjusted factor cost prices would have been desirable to use to weight product groups, branches or end use categories in measuring changes in Cuba's industrial production potential. I was able, however, to directly compare the growth bias of Cuban and Guatemalan prices, and in some cases the comparison was also made with 1967 Peruvian and 1972 US prices.

The best estimate of this study is that real annual Cuban industrial growth during 1965–84 was 6.3%. This is a very healthy, if not impressive, rate of growth and stands out in sharp relief when compared to the growth experience in the rest of Latin America. It remains to understand the reasons for this favorable industrial performance. Some will attribute it to high investment ratios, propitious structural change, heavy human capital investments, full resource employment or central planning, others will argue that Soviet aid is responsible and yet others will seek more complex explanations.[24] Such an analysis, however, will have to await a future study.

NOTES

1. ECLA (1985), p. 8; *Latin America Weekly Report* (3 January 1986, p. 8). ECLA is the Economic Commission for Latin America of the United Nations.

2. Cuban National Income or Net Material Product (which is based on value added and in theory eliminates double counting) grew even more rapidly according to

official statistics, at 7.9% per capita per annum during 1980–85. See CEE (1984) and Castro's speech to the Third Party Congress in February 1986. My calculation of 7.9% is based on the conservative assumption that GSP and NMP expanded at the same rate, 4.9%, in 1985.

3. The actual range of techniques is rather broad. Mongrel and makeshift methods are also prevalent because of the frequent irregularity and paucity of data. Some of the more important studies are: Becker (1969); Bergson (1950; 1961); Converse (1982); Greenslade (1976); Hodgman (1954); Kaplan and Moorsteen (1960); Montias (1967); Pitzer (1982); Staller (1962); and Treml and Hardt (1972).

4. Marer (1985a; 1985b).

5. Mesa-Lago and Pérez-López (1982).

6. For a critique of this study see Brundenius and Zimbalist (1985a; 1985b).

7. Pérez-López (1983; 1985).

8. CIA (1984).

9. Domínguez (1985), p. 103.

10. CEE (1984), pp. 89–91.

11. That is, following the findings of Grossman (1960), Nove (1977, Chap. 13) and others, WEFA accepts the official physical production data as reliable.

12. Since WEFA sectoral weights are from a 1974 conversion study by the Cuban State Statistical Committee and 1974 is the base year, the branch GVO distribution in 1975 was used to proxy for the same in 1974. Cuba changed its branch classification system in 1977 and has published consistent series for the new system only back to 1975. See CEE (1977), Chap. 6.

13. Brundenius and Zimbalist (1985b).

14. Disaggregated information on individual products is available for some branches in Brundenius and Zimbalist (1985a; 1985b).

15. In reference to the Soviet economy these prices (gross of turnover tax) are generally called "industrial prices." I shall use the term "producer prices," however, as it is a direct translation from the Spanish.

16. Among other things, to sustain the WEFA finding of negative industrial growth in 1981 it would entail rejecting the "Converse Principle." In his recent study of Soviet industrial growth, Ray Converse writes (1982; p. 201): "We expect that even though the official indexes are biased, they nonetheless reflect actual growth trends."

17. See Bergson (1961) for a full discussion of this phenomenon.

18. Conversations with officials at the Cuban State Statistical Committee in October 1985 and March 1986.

19. My source for the 1967 Peruvian industrial prices is Ministerio de Indústria y Comercio (1969). I also attempted to procure US wholesale prices for the 206 products listed in the *Anuario Estadístico de Cuba*. The US Bureau of Labor Statistics publication *Producer Prices and Price Indexes* contains only a limited number of actual producer prices. (I did obtain sufficient price data from the December 1972 issue of this publication to estimate growth in four branches. In three cases the estimate with US prices was higher than with Cuban prices. In the fourth case the estimate with US prices was above the estimate with Peruvian prices.) For the vast bulk of commodities only price indexes, not actual prices, are available.

I spoke with several people in the producer price office of the BLS in an effort to obtain the relevant prices. With apologies, however, I was at each instance turned away with the explanation that the price data are proprietary, i.e., the BLS promises the companies not to release it. Nor could I persuade the BLS to give me price ratios among groups of products. It is an interesting irony that it is easier to obtain price information in a centrally-planned economy than in a market economy.

20. For discussions of the new product pricing issue see, *inter alia*: Greenslade (1972, pp. 181–186) and Pitzer (1982, p. 13).

21. Brundenius (1985).

22. Calculated from Brundenius (1985), Table 1, Appendix.

23. Cohn (1972) considers different but overlapping categories from those in the present study in analyzing the sources of difference in Western and official estimates of Soviet national income growth.

24. I have argued elsewhere that the positive impact of generous Soviet price subsidies and balance of payments aid must be analyzed in the context of the US blockade and that the CIA estimates regarding the magnitude of this aid are seriously flawed and exaggerated. See Zimbalist (1982) and Radell (1983).

REFERENCES

Becker, Abraham, *Soviet National Income, 1958–1964* (Berkeley: University of California Press, 1969).
Bergson, Abram, "Soviet national income and product in 1937," *Quarterly Journal of Economics*, Vol. 64 (1950), pp. 208–241, 408–441.
Bergson, Abram, *The Real National Income of Soviet*

Russia since 1928 (Cambridge: Harvard University Press, 1961).

Brundenius, Claes, "The role of capital goods production in the economic development of Cuba," Paper presented to conference of *Technology Policies for Development* (Lund, Sweden: Research Policy Institute, University of Lund, 29–31 May 1985).

Brundenius, Claes, and Andrew Zimbalist, "Recent studies on Cuban economic growth: A review," *Comparative Economic Studies*, Vol. 27, No. 1 (Spring 1985a), pp. 21–45.

Brundenius, Claes, and Andrew Zimbalist, "Cuban economic growth one more time: A response to 'Imbroglios'," *Comparative Economic Studies*, Vol. 27, No. 3 (Fall 1985b), pp. 115–131.

Central Intelligence Agency, *The Cuban Economy: A Statistical Review* (Washington, DC: National Technical Information Service, 1984).

Cohn, Stanley, "National income growth statistics," in Treml and Hardt (1972).

Comité Estatal de Estadísticas, *Anuario Estadístico de Cuba* (Havanna, Cuba, 1973 through 1984).

Converse, Ray, "An index of industrial production in the USSR," Joint Economic Commission (1982).

Domínguez, Jorge, "Cuba: charismatic communism," *Problems of Communism* (Sept–Oct 1985), pp. 102–107.

Economic Commission for Latin America (ECLA), *Cepal Review 25* (April 1985).

Greenslade, Rush, "Industrial production statistics in the USSR," in Treml and Hardt (1972).

Greenslade, Rush, "The real gross national product of the USSR, 1950–75," in Joint Economic Commission (1976).

Grossman, Gregory, *Soviet Statistics of Physical Output of Industrial Commodities: Their Compilation and Quality* (Princeton: Princeton University Press, 1960).

Hodgman, Donald, *Soviet Industrial Production, 1928–1951* (Cambridge: Harvard University Press, 1954).

Joint Economic Commission, *Soviet Economy in a New Perspective* (Washington, DC: US Government Printing Office, 1976).

Joint Economic Commission, *USSR: Measures of Economic and Development*, 1950–80 (Washington, DC: US Government Printing Office, 1982).

Joint Economic Commission, *East European Economies: Slow Growth in the 1980s* (Washington, DC: US Government Printing Office, 1985).

Kaplan, Norman, and Richard Moorsteen, "An index of Soviet industrial output," *American Economic Review*, Vol. 50 (June 1960).

Marer, Paul, "Alternative estimates of the dollar GNP and growth rates of the CMEA countries," in Joint Economic Commission (1985).

Marer, Paul, *Dollar GNPs and Growth Rates of the USSR and Eastern Europe* (Baltimore: The Johns Hopkins University Press for the World Bank, 1985).

Mesa-Lago, Carmelo, and Jorge Pérez-López, "Study of Cuba's MPS, its conversion to SNA, and estimation of GDP/capita and growth rates," Second Workshop on CPE National Income Statistics (Washington, DC: The World Bank, October 1982).

Mesa-Lago, Carmelo, and Jorge Pérez-López, "Imbroglios on the Cuban economy: A reply to Brundenius and Zimbalist," *Comparative Economic Studies*, Vol. 27, No. 1 (Spring 1985), pp. 47–83.

Ministerio de Indústria y Comercio, *Estadística Industrial*, 1967, No. 21 (Lima, Peru: 1969).

Montias, John M., *Economic Development in Communist Rumania* (Cambridge: MIT Press, 1967).

Nove, Alec, *The Soviet Economic System* (London: George Allen and Unwin, 1977).

Pérez-López, Jorge, *Construction of Cuban Economic Activity and Trade Indexes* (Washington, DC: Wharton Econometrics Forecasting Associates, November 1983).

Pérez-López, Jorge, "Real economic growth in Cuba, 1965–82," Paper presented at the Latin America Scholars Association Meetings (Albuquerque, New Mexico: April 1985).

Pitzer, John, "Gross national product of the USSR, 1950–80," Joint Economic Commission (1982), pp. 3–160.

Radell, Willard, "Cuba–Soviet sugar trade, 1960–1976," *The Journal of Developing Areas*, Vol. 17, No. 3 (April 1983), pp. 365–379.

Staller, George, "Czechoslovak industrial growth: 1948–1959," *American Economic Review*, Vol. 52, No. 3 (1962), pp. 384–407.

Treml, Vladimir, and John Hardt (Eds.), *Soviet Economic Statistics* (Durham: Duke University Press, 1972).

Zimbalist, Andrew, "Soviet aid, U.S. blockade and the Cuban economy," *Comparative Economic Studies*, Vol. 24, No. 4 (Winter 1982), pp. 137–145.

·6·

Development and Prospects of Capital Goods Production in Revolutionary Cuba

CLAES BRUNDENIUS

1. CAPITAL GOODS PRODUCTION AND INDUSTRIALIZATION

Capital goods (usually defined as machinery and transport equipment[1]) have played an important role in the capital formation and the industrialization process in the leading industrial nations of today. This has been true of the nations that started to industrialize in the 19th century as well as of the relative newcomers such as Japan and the Soviet Union.[2] The capital goods sector has been crucial not only in terms of its linkage effects with the rest of the economy, but also because it has acted as a decisive instrument for the generation and diffusion of technological change throughout the economy. This is so because there is an important learning process involved in capital goods production with a high degree of specialization that is conducive not only to an effective learning process but also an effective application of what is being learned.[3]

The relatively high level of technology usually required for the production of capital goods, often involving specialized skills, has represented an important barrier to the entry into these industries by developing countries, especially those with a weak industrial base and a lack of skilled people. Another obstacle is the economies of scale that are often required for efficient capital goods production. Most of the developing countries today have only embryonic capital goods industries or no such industries at all. In fact, 74% of capital goods production in the developing countries today is concentrated in six or seven Newly Industrializing Countries (NICs).[4] But the developing countries are large importers of capital goods. According to a UNIDO study the developing countries share in capital goods output in 1980 was only 6% of world output while their share of world imports of capital goods was 29%.[5]

The absence of a domestic capital goods industry is undeniably a serious hindrance to rapid capital accumulation and a sustained rate of economic growth for a developing country. Such a country could, of course, for quite some time depend on the importation of such capital goods but then it would also be forced to depend on the traditional export of primary goods, the price prospects of which are rather gloomy.

This article addresses the experience of Cuba, a country with a socialist economy, which has faced an additional problem. At the outset of the Revolution in 1959, Cuba emphasized the meeting of basic needs and the rapid elimination of unemployment as relatively *short-term* goals. At the same time, it was more or less taken for granted that the needs of capital accumulation — a *sine qua non* for the sustained rate of growth of the economy in the long run — would not be neglected. As the experiences of other countries have demonstrated, however, it is quite difficult to reconcile these two objectives in the short run, and Cuba proved to be no exception, as will be discussed below.

There are both costs and benefits to building up an indigenous capital goods industry. There are immediate benefits of the process such as saving foreign exchange, making the country less vulnerable to price changes of capital goods on the world market and assuring secure supplies of spare parts. More long-term benefits may be experienced when skill acquisition and product innovations in the capital goods sector lead to productivity increases in the rest of the economy as well as in the capital goods sector itself.

But there may also be large *social costs* in building up an indigenous capital goods sector. These costs may be reflected in a lower efficiency of locally produced capital goods, as compared with similar goods on the world market. This might lead to a diversion and waste of resources. The *timing* of the building up of an indigenous capital goods sector is thus essential in planning. The planning of capital goods production is therefore a long-term operation. The sector's diversity extends from the production of small parts and components to large and sophisticated machinery. Programming and planning thus embrace not only production of machinery which is economically viable, but also evaluation of various alternative production possibilities in terms of socioeconomic costs and benefits.

Machinery and equipment have always been considered important in a "socialist" growth strategy. Such a strategy is often based on two postulates: that industrialization should take place at the fastest rate possible, and that capital goods growth should continuously outpace that of consumer goods. This latter postulate is sometimes referred to as "the law of the preferential expansion of Department I," with a reference to Marx's celebrated model of accumulation. Department I in Marx's analysis is, however, a much wider category since it includes also what we today call intermediate goods (such as steel, paper, cement and fertilizers) and not only means of production in the sense of machinery and equipment. But even so it is quite clear that also in Marx's analysis the "law" implies a preferential increase in the production of capital goods (machinery and equipment).

Marx's model was in the forefront in the industrialization debate in the Soviet Union in the 1920s as well as in the discussion on long-term planning in India in the 1950s. Since I have dealt with these experiences at some length in other contexts,[6] it might be sufficient here to stress that for Feldman in the Soviet Union as well as for Mahalanobis in India (and later on Raj and Sen) the fraction of investment allocated to the investment goods sector itself (and hence machinery and equipment) was the key element in the growth model of economy.

One major flaw in the Feldman-Mahalanobis model was, however, that it excluded foreign trade and was based — at least implicitly — on the assumption of a closed economy. This is a highly unrealistic assumption today and could at best be valid for large economies with huge potential markets and with large untapped reserves of natural resources. The Raj-Sen model is more realistic since it allows for "stagnant foreign exchange earnings." The model demonstrates how such a rigid exchange constraint — rather typical for most non-oil exporting developing countries today — justifies the building up of a domestic capital goods industry. If limited foreign exchange is used to import consumer goods, then the economy will *stagnate*, yet if it is used to import capital goods for the production of consumer goods, there will be a steady rate of investment and a *steady* rate of growth. If, however, foreign exchange is used to import capital goods to expand the capital goods sector itself then the capital goods sector will increase at an *accelerated* rate, with a rise in the investment ratio as a result, and, consequently, the economy as a whole will tend to grow at an accelerated rate.

2. THE STRATEGY AND EXPERIENCE OF CUBA

(a) *Starting from scratch*

On the eve of the Revolution, in 1958, Cuba was basically a sugar economy. A slow but clear attempt at industrialization had been taking place since the end of World War II, but this proved to be a difficult process since it met with much resistance from powerful sectors of the Cuban bourgeoisie. It is clear from reading some recently released documents from that period[7] that there were strong groups (particularly plantation owners) with a vested interest in Cuban adherance to the Reciprocity Treaty with the United States which gave Cuban sugar preferential treatment on the United States market in return for tariff exemption on US industrial exports to the island. The National Association of Manufacturers (of Cuba) does not appear to have been strong enough to counter the vested interest in perpetuating Cuba's dependency on the United States with its retarding effects on the industrialization of the country.

It has been estimated that at the end of 1950s, on the eve of the Revolution, sugar accounted for about one-fourth of total industrial output while the whole metallurgical sector (including the mechanical industry) only represented 1.4% of industrial output.[8] The production of the engineering goods industries (ISIC 38) in the 1950s was characterized by small workshops, of an artisan or semi-artisan kind, with low levels of mechanization, practically no engineers and insufficient skills among workers.[9]

The obsolescence of the Cuban sugar industry was disastrous; the most "modern" plant dated from 1927, and 40% of the *centrales* (sugar mills) had been built during the 19th century.[10] This obsolescence meant that most of the machinery, equipment and spare parts for this industry were no longer produced on the world market and therefore had to be produced nationally in order to guarantee continuing supplies.

When the revolutionary government took power in January 1959, capital goods production was not on the agenda. The main preoccupation of the government was to tend to the *immediate* social needs of the people by eradicating illiteracy, providing health care, reducing the tremendously high unemployment rates and starting redistributive reforms (for instance, rent legislation and the agrarian reform).[11] It was more or less taken for granted that capital accumulation would occur, almost automatically, at a later stage.

Industrial production increased during the first years of the Revolution but mainly as a result of utilization of idle capacity. Several well-known economists visited the island in those first years and they were all very optimistic about the prospects for the rapid industrialization of Cuba. It was in this mood that Michael Kalecki, the well-known Polish economist and planner, in 1960 sketched a Five-Year Plan for the period 1961–65, basing his optimistic projections on demand criteria only (as a result of the income redistribution effects after the Revolution) and seriously neglecting supply conditions.[12] Kalecki projected an *annual* rate of growth of the Cuban economy of 13%, spearheaded by a spectacular growth of the capital goods sector of no less than 36% per year! Now it should be pointed out that capital goods industries only accounted for an estimated 2.9% of total industrial output at the beginning of the plan period (see Table 1), even so, Kalecki foresaw the construction of a wide range of capital goods industries within such a short span of time.

In spite of mounting pressures against resources which were being felt already in 1961, the Four-Year Plan (1962–65) presented by the government in the fall of 1961 contained much of the flavor of Kalecki's original sketch. At the center of this "first growth strategy" was a rapid industrialization to be achieved through a chain of important-substituting industries including metallurgy, chemical products, machinery and even an assembly plant for passenger cars. Most of the machinery and equipment for these new plants was to be supplied by the Soviet Union and other socialist countries. Nowhere in the plan was it envisaged that there might be curtailment of consumption via rationing as a consequence of this high investment rate, because there was an underlying assumption that was more or less taken for granted: the rapid expansion of non-sugar agriculture would lead to self-sufficiency in food production and eliminate queuing and problems of food supply by the end of 1962.[13]

Harsh reality very soon made these plans unworkable. Far from achieving an accelerated growth, the Cuban economy encountered more and more difficulties, especially after the United States announced a full-scale naval blockade against the island following the "missile crisis" in October 1962. The plan was soon abandoned altogether. Already in October 1960 President Eisenhower had imposed an embargo on all exports to Cuba, effectively curtailing vital Cuban imports of raw materials and spare parts for its industry. In February 1962 President Kennedy extended the embargo to all imports from Cuba, those coming directly from Cuba as well as via a third country and ships going to Cuba were blacklisted and denied entry into US ports.[14] Surrounded then by hostile, ex-trade partners, Cuba had to resort to foreign financing of its necessary imports of machinery and equipment (and even basic consumer goods), while, at the same time, measures were taken to guarantee a decent living standard to the majority of the population.

But even if economic growth in Cuba was faltering in the 1960s, this does not mean that there was no increase in capital goods production. On the contrary, the capital goods industry was the most rapidly expanding sector in the Cuban economy during the second half of the 1960s. In those years there was also a drastic restructuring of imports, with priority given to the import of machinery and equipment, especially those needed to increase worker productivity and land yields in agriculture.

In 1964 Ernesto Ché Guevara, then Minister of Industry, inaugurated a mechanical plant in Santa Clara, the output of which was to supply the industries related to agriculture with spare parts and also, in the longer run, machinery and equipment. This is when Cuba's long, burdensome (but not always impatient) process of "learning by doing" began. The embargo imposed by the United States actually converted such a "learning process" from a virtue into a necessity. Shoemakers and carpenters were retrained to become skilled industrial workers, manufacturing spare parts for the old, obsolescent machine park of Cuban industry. When this process started in 1964, the capital goods industry in Cuba only employed between one and two thousand people. The Santa Clara plant was the largest with 350 workers.[15]

The 1960s were thus a more difficult period in Cuba. It has been estimated that real GDP grew at a rate of only 1.5% between 1961 and 1970, or just about making up for the increase in population.[16] But there were positive trends as well. The massive effort to mobilize resources for education and health and the gradual elimination of the specter of unemployment were important instruments in creating a well educated, healthy and motivated work force in Cuba, an important condition for the rapid economic transformation of the country.

Already in the 1960s there had been a modest, but nevertheless important, increase in non-sugar manufacturing activities with an annual rate of growth of 4.1% between 1961 and 1970. Engineering goods broadly defined (ISIC 38) increased at an annual rate of 6.7% (see Table 1), far from the Kalecki dreams of 36% per year but nevertheless a good start. But even so, engineering goods still accounted for only 3.6% of total industrial output as of 1970.

(b) *The first Five-Year Plan* (1976–80)

During the first half of the following decade (1971–75) the Cuban economy experienced a real boom. Real GDP is estimated to have increased at an unprecedented rate of 7.8% per year. Non-sugar manufacturing grew by 10.8% per year and the engineering goods industries by no less than 24.5% per year (see Table 1). One likely explanation for this boom is that much of the construction work that was started in the late 1960s was completed during this period. The construction sector grew at a rate of 22.6% during the first half of the 1970s. Gross investments again started to increase after having stagnated in the 1960s. Gross investments reached the amount of 2.3 billion pesos by 1975, or almost three times the 1970 level (see Table 2). Gross investments as share of the Global Social Product (GSP)[17] grew from a record low of 9.6% in 1970 to 16.4% in 1975.

Table 1. *Gross value of production of engineering goods industries and total industry in constant* (1981) *prices* (million pesos)

	Total industrial output*		Output of engineering goods industries (ISIC 38)		Engineering goods share of total industry (%)
	Value	Index	Value	Index	
1961	2,567.8	46.6	73.4	18.6	2.9
1962	2,623.0	47.6	75.7	19.2	2.9
1963	2,584.4	46.9	74.8	19.0	2.9
1964	2,689.1	48.8	76.6	19.5	2.8
1965	2,793.8	50.7	68.2	17.3	2.4
1966	2,744.2	49.8	72.7	18.3	2.6
1967	3,041.7	55.2	76.6	19.5	2.5
1968	2,992.1	54.3	94.3	23.9	3.2
1969	2,959.1	53.7	87.4	22.2	3.0
1970	3,609.3	65.5	131.7	33.4	3.6
1971	3,818.7	69.3	160.0	40.6	4.2
1972	4,055.7	73.6	185.6	47.1	4.6
1973	4,540.6	82.4	249.7	63.4	5.5
1974	4,898.7	88.9	310.4	78.8	6.3
1975	5,510.4	100.0	393.8	100.0	7.1
1976	5,582.9	101.3	420.4	106.8	7.5
1977	5,694.8	103.3	445.5	113.1	7.8
1978	6,214.9	112.8	535.3	135.7	8.6
1979	6,392.7	116.0	616.9	156.7	9.7
1980	6,583.6	119.5	642.0	163.0	9.8
1981	7,729.2	140.3	795.4	202.0	10.3
1982	8,043.0	146.0	889.6	225.9	12.1
1983	8,466.3	153.6	1,051.4	267.0	12.4
1984	9,377.4	170.2	1,224.6	311.0	13.1
1985†	9,996.7	181.4	1,321.9	335.7	13.2

Source: Total industrial output: Constructed from series given in 1981 prices for 1980–84 in *Anuario* (1981, 1984), in 1978 prices for 1975–80 in *Anuario* (1982) and 1965 prices in Brundenius (1984), Tables A.2.7. and A.2.9. Output of engineering goods industries: Brundenius (1986), Appendix Table 1.
1985 figures are estimates by the author using trend as of November that year given in *Boletín Estadístico Mensual* (November 1985).
*Includes mining and electricity.
†Estimates based on first 11 months of 1985.

The economic strategy behind the boom in the 1970s had been formulated already in the 1960s. Miguel Figueras, until April 1986 Vice-Minister of Planning, and in the 1960s a close aide of Ché Guevara, gives an interesting account of how the strategy was gradually hammered out.[18]

It was clear, says Figueras, that one of the objectives of the blockade imposed by the United States was to paralyze the Cuban economy (by preventing the flow of spare parts to US made machinery which accounted for 80–90% of the total machine park). In view of this situation, the

Table 2. *Gross investments and global social product*, 1963–85 (million pesos, current prices)

	Gross investments	Global social product	GI/GSP (%)
1963	732.6	6,013.2	12.2
1964	809.5	6,454.5	12.5
1965	841.7	6,770.9	12.4
1966	935.1	6,709.3	13.9
1967	1,031.5	7,211.6	14.3
1968	918.0	7,330.9	12.5
1969	896.4	7,236.1	12.4
1970	800.1	8,356.0	9.6
1971	963.8	8,966.5	10.7
1972	1,094.0	10,417.9	10.5
1973	1,475.0	11,921.8	12.4
1974	1,711.9	13,149.0	13.0
1975	2,304.2	14,063.4	16.4
1976	2,587.4	14,458.2	17.9
1977	2,765.9	14,772.8	18.7
1978	2,623.6	16,457.6	15.9
1979	2,605.8	16,986.8	15.3
1980	2,739.1	17,605.6	15.6
1981	3,386.1	22,172.5	15.3
1982	2,995.2	23,112.8	13.0
1983	3,408.5	24,336.9	14.0
1984	3,989.4	26,104.9	15.3
1985*	4,085.1	27,410.1	14.9

Source: *Boletín Estadístico de Cuba* (1968, 1971); *Anuario* (1974, 1979, 1981, 1982, 1984); *Boletín Estadístico Mensual* (November 1985); *Cuba en Cifras* (1981); *Cuba: Cifras Estadísticas* (1981).
*Estimate based on first 11 months of 1985.

revolutionary government drew up the following strategy:
1. To attempt to find equivalent spare parts and components in the socialist countries and in the case where such goods were not available, to get assistance agreements with those countries for the production of such goods specifically for Cuba.
2. To try to purchase some similar spare parts from some capitalist countries maintaining relations with Cuba. This proved to be extremely difficult for two reasons: because of the lack of foreign exchange (in convertible currencies) and because of the pressure exercised by the United States on Western countries not to trade with Cuba.
3. To develop a massive national movement for the production of spare parts and to strengthen inter-plant collaboration in order to find joint solutions. This should be accomplished by:
   (i) having the workshops in each plant fabricate (or find out how to fabricate) the greatest number of spare parts and attempt to fabricate spare parts for other plants less skilled in this field;
   (ii) creating specialized factories for the production of spare parts, training and upgrading workers on the job, acquiring the necessary machine tools from the socialist countries and requesting the assistance of technical expertise from these countries in training and, if necessary, supervising production;
   (iii) asking the workers and technicians in the few mechanical industries that existed at the time, most of them supplying the sugar industry, to increase production and to try to incorporate new types of components and equipment in their lines of production.

That was the recipe followed and, says Figueras, it made the US embargo fail. In view of the accelerated growth of the economy during the first half of the 1970s, the projections for the first Five-Year Plan (1976–80) were also cautiously optimistic. JUCEPLAN (the planning agency) prepared three variants of the plan to allow for possible reverses in the price of sugar (which in 1974 had skyrocketed to unprecedented levels). Since prices fell more than expected (much more than even the most pessimistic variant of the plan had foreseen), 22 major investment projects had to be cancelled, most of which were to be financed with credit lines from the West.[19] Thanks to steadily improving terms of trade with the Soviet Union, the Cuban trade balance with that country was favorable in 1976 for the first time since the early 1960s.

The first Five-Year Plan thus projected continuing accelerated growth and the termination of a number of ambitious industrial projects. Among the industrial projects that were planned and actually executed during this period were two cement plants with an annual capacity of three million tons, two medium-sized sugar mills (the first ones to be built since 1927!), four bulk sugar terminals, one modern textile plant, two piping and irrigation spray units factories, one plant for wheels for heavy equipment and the start of the modernization of the nickel refining and petroleum refining plants.[20] It should be mentioned that in the case of the sugar factories all the technical plans were designed in Cuba and more than 60% of the machinery, including the mechanized and automated units, were of Cuban origin.[21] The most successful part of the Cuban expansion program for its capital goods sector has no doubt been the development of equipment for the sugar industry (see below).

The growth rate of the economy during the first Five-Year Plan was considerably lower than expected. National income[22] grew at a rate of only 3.0% (compared with the originally planned rate of 6%), agriculture by 0.4% and industry by 1.9% (see Table 3). On the other hand, the investment goods industries grew considerably. The construction sector, for instance, grew at a rate of 4.6% and the gross value of output of the engineering goods industries increased by 10.3% per annum.

During the first Five-Year Plan the share of engineering goods in total industrial output increased from 7.1% in 1976 to 9.8% in 1980 (Table 1). An increasing share of investment also went to the engineering goods industries as well as to industrial activities as a whole (see Table 4).

(c) *The second Five-Year Plan* (1981–85)

When the guidelines for the second Five-Year Plan (1981–85) were drawn up in 1980 the prospects for its implementation seemed poor.

The last year of the first plan (1980) had ended with a considerable slowing down of growth rates on practically all fronts and targets had not been fulfilled in a number of areas. The growth rates for the second plan were accordingly set more realistically with an overall rate of growth of the economy of 5% per annum.[23] However, it was considered that the capital goods sector should be instrumental in making this growth target realistic. The capital goods industries were — according to the Plan — expected to increase at an impressive rate of 12 to 15% per year. The detailed plan for these industries spelled out these ambitions in the following terms:

— To start, in the five-year period, the construction of no less than seven new sugar refineries; to extend 23 and modernize 18 of the existing ones, to reach a potential daily granulated capacity of about 690,000 tons of cane in 1985.

— To continue the development of engineering production for the sugar industry to meet the growth plans of the latter. To reduce the use of imported equipment and

Table 3. *Growth rates during the Five-Year Plans*

	National income	Agriculture†	Construction†	Industry†	Engineering goods industry‡	Gross investments
First Plan (1976–80)						
1976	5.1	3.1	5.4	4.4	6.8	12.3
1977	8.3	4.8	10.6	0.7	6.0	6.9
1978	6.9	7.6	4.8	8.6	20.2	−5.1
1979	0.7	−20.0	−11.6	−0.4	15.2	−0.7
1980	−5.2	−10.2	13.8	−3.1	4.1	5.1
Average 1976–80	3.0	0.4	4.2	1.9	10.3	3.5
Second Plan (1981–85)						
1981	20.8	18.7	25.1	29.8	23.9	23.6
1982	5.0	−5.2	1.1	6.2	11.8	11.5
1983	5.5	−8.5	14.2	7.1	18.2	13.8
1984	7.4	2.7	19.2	13.4	16.5	17.0
1985§	4.8	4.2	2.7	6.6	7.9	2.4
Average 1981–85	8.6	1.9	12.1	12.3	15.6	8.3

Source: *Anuario* (1984), Table III.5 and *Beletín Estadístico Mensual* (November 1985).
*National Income (or Net Material Product) equals gross value of production minus productive consumption in the productive sphere (that is, agriculture, industry, construction, transport, communication and trade).
†Net value of production.
‡Gross value of production.
§Estimate based on first 11 months of 1985.

Table 4. *Gross investments in agriculture and industry, 1974–84*
(million pesos, current prices)

	Total economy Value	%	Agriculture Value	%	Industry Value	%	Engineering industry Value	%
1974	1,711.9	100.0	466.3	27.2	328.2	19.2	NA	NA
1975	2,304.2	100.0	559.6	24.3	630.1	27.3	63.2	2.7
1976	2,587.4	100.0	553.9	21.4	807.9	31.2	82.9	3.2
1977	2,756.9	100.0	540.1	19.5	966.5	34.9	93.7	3.4
1978	2,634.6	100.0	466.0	17.8	987.1	34.2	97.7	3.7
1979	2,605.8	100.0	435.5	16.7	1,003.7	38.5	148.0	5.7
1980	2,739.1	100.0	584.5	21.3	1,007.1	36.8	145.0	5.3
1981	3,386.1	100.0	895.5	26.4	1,176.6	34.7	142.6	4.2
1982	2,995.2	100.0	737.1	24.6	1,058.5	35.3	143.5	4.8
1983	3,408.5	100.0	760.5	22.3	1,244.0	36.5	125.5	3.7
1984	3,989.4	100.0	875.8	22.0	1,422.5	35.7	135.3	3.4

Source: *Anuario* (1981, 1982, 1983 and 1984).

components and in addition to produce these items for export.
— To achieve sustained growth in the production of the non-electrical machinery and metal products industries in order to meet the demand more fully. To eliminate imports and increase participation in exporting.
— To increase the production of agricultural machinery, equipment and implements-. . . to develop new lines with a view to guaranteeing to a larger extent the internal demand and the replacement of imports.
— To increase and develop the production of new lines of transport equipment, to work on the production of diesel engines on an experimental scale and to evaluate the results.
— To increase the production of machinery and equipment for the sugar industry, including substantial parts of sugar refineries with the aim of guaranteeing the national necessities and to convert this production into one of the fundamental export lines of the engineering industry.
— To increase and develop the production of steel structures and technological equipment, including complete lines and plants, to increase the participation of the "equipment" content from national production in the investments to be made in the period, and developing exports.
— To increase those productions which support constructional activities such as moulds for concrete building elements, bronze fittings, locks and accessories for, *inter alia*, sanitary furniture.

— To raise the utilization of production capacities for stainless steel equipment, parts and spares.
— To develop the production of fishing and leisure craft, for both export and for replacement, with the introduction of more productive technologies . . . improving both quality and design, according to end-use and destination.
— To encourage the production of technical means for computing, with a view to contributing to national necessities and to creating sources of exportable funds.[24]

As seen from this listing of medium-term goals, during the plan period, the Cuban capital goods industry would continue to specialize in the construction of machinery, equipment and spare parts, including the production of complete plants.

At the same time priority would

continue to be given to the creation of a solid base for the development, design, planning and manufacture of new types and models of equipment for automobile transport (both freight and passenger) and for the railways, including technical and feasibility studies, technical projects, a greater part of the equipment (60% or more), with the construction, erection and commissioning of the said installations with the object of continuing to replace imports of those products which can be manufactured in the country and also to create new and exportable items.[25]

The plan period could not have gotten off to a better start. In 1981 there was a tremendous boom in the economy with two-digit growth rates (see Table 3). Even if growth rates since then tapered off somewhat there seems to be little doubt that most of the overall objectives of the plan were fulfilled. Quite an achievement considering that the rest of Latin America in the same period was confronting its worst economic crisis since the Depression. By the end of the plan period Cuba produced 1.3 billion pesos worth of engineering goods, accounting for 13.2% of total industrial output (see Table 1), a doubling of what production had been in 1980.

3. THE CAPITAL GOODS INDUSTRIES

Let us now take a closer look at the capital goods industries in Cuba. As mentioned before[26] it is not always so easy to separate purely capital goods (machinery and equipment) from engineering goods in general (which also include consumer durables). Cuban statistics, however, allow for the distinction between non-electrical machinery, electrical machinery and metal products, groups that more or less correspond to ISIC (United Nations International Standard Industrial Classification of All Economic Activities), namely ISIC 381 (fabricated metal products), ISIC 382 (manufacture, except electrical), ISIC 383 (manufacture of electrical machinery and appliances) and ISIC 384 (manufacture of transport equipment). Such a breakdown is available for Cuba after 1975 and production figures for the three subsectors are shown in Table 5.

As can be seen from Table 5 the most significant increase has taken place in the non-electrical machinery sector (which in Cuban statistics includes transport equipment) with an almost fourfold production increase between 1975 and 1985. The electrical machinery sector experienced in the same period a threefold increase while the metal products sector more than doubled its output. All three subsectors accelerated output growth during the second Plan (1981–85). Non-electrical machinery went from a rate of 11.4% per year during the first plan to a rate of 17.3% per year during the second plan, electrical machinery from 8.3% during the first plan to 15.1% during the second plan, and metal products from 6.6% during the first plan to 10.6% during the second.

A closer look at the subsectors reveals that non-electrical machinery is by far the most important of the engineering goods industries, accounting for 7.2% of total industrial output in 1981 and 9.3% in 1984 (see Table 6). In terms of employment, the branch is even more significant with 12.2% of total industrial employment in 1981 and 12.4% in 1984. The contribution of electrical machinery and metal products is much more modest both in terms of output and employment between 3% to 4% together in both 1981 and 1984).

Labor productivity is considerably lower in the engineering goods industry than in industry as a whole, although the gap narrowed significantly between 1981 and 1984 (Table 6). Productivity is highest in the metal products sector which is only

Table 5. *Gross value of production of engineering goods industries, 1975–85* (million pesos, 1981 prices)

	Non-electrical machinery* (ISIC 382 & 384) Value	Index	Electrical machinery (ISIC 383) Value	Index	Metal products (ISIC 381) Value	Index	Total (ISIC 38) Value	Index
1975	253.7	100.0	50.7	100.0	89.4	100.0	393.8	100.0
1976	275.0	108.4	52.4	103.4	93.0	104.0	420.4	106.8
1977	292.4	115.3	55.9	110.3	97.2	108.7	445.5	113.1
1978	359.2	141.6	64.0	126.2	112.1	125.4	535.3	135.9
1979	420.1	165.6	78.3	154.4	118.5	132.6	616.9	156.9
1980	434.4	171.2	75.4	148.7	123.2	137.8	642.0	163.0
1981	553.4	218.1	100.6	198.4	141.4	158.2	795.4	202.0
1982	642.7	253.3	88.5	174.6	158.4	177.2	889.6	225.9
1983	746.0	294.0	119.4	235.5	186.0	208.0	1,051.4	267.0
1984	876.2	345.4	144.5	285.0	203.9	228.1	1,224.6	311.0
1985†	866.4	380.9	152.0	299.8	203.5	227.6	1,321.9	335.7

Source: *Anuario* (1981, 1982, 1983 and 1984); *Boletín Estadístico Mensual* (November 1985).
*Including transport equipment.
†Estimate based on first 11 months of 1985.

Table 6. *Engineering goods industries: Basic indicators 1981 and 1984*

	Non-electrical machinery (ISIC 382 & 384)	Electrical machinery (ISIC 383)	Metal products (ISIC 381)	Engineering goods industry (ISIC 38)	Total industry
1981					
Gross value of production (MMP)*	553.4	100.6	141.4	795.4	7,729.2
Number employed (000)	70.5	12.3	11.2	94.0	576.4
Productivity†	7,849.0	8,178.0	12,625.0	8,462.0	13,409.0
Average salary/year	2,220.0	2,040.0	2,184.0	2,196.0	2,136.0
Gross investments (MMP)	112.9	19.7	10.0	142.6	1,176.6
Capital invested (MMP)‡	350.7	71.0	59.2	480.9	6,889.9
Investment ratio	0.20	0.20	0.07	0.18	0.15
Capital/output ratio	0.63	0.71	0.42	0.60	0.89
1984					
Gross value of production (MMP)*	876.2	144.5	203.9	1,224.6	9,377.4
Number of employed (000)	82.8	14.2	14.6	111.6	665.7
Productivity†	10,582.0	10,176.0	13,966.0	10,973.0	14,087.0
Average salary/month	2,460.0	2,256.0	2,412.0	2,428.0	2,328.0
Gross investments (MMP)	103.7	12.6	19.0	135.3	1,422.5
Capital invested (MMP)‡	544.9	79.1	87.9	711.9	8,788.0
Investment ratio	0.12	0.09	0.09	0.11	0.15
Capital/output ratio	0.62	0.55	0.43	0.58	0.94

Source: *Anuario* (1984) and data supplied to the author by CEE (Comité Estatal de Estadísticas).
*Million pesos.
†Gross value of production per employed (pesos).
‡At year end.

slightly below the industrial average, with non-electrical machinery falling far behind in 1981. The non-electrical machinery sector, however, seems to be the most dynamic one and by 1984 the productivity gap had narrowed considerably. Between 1981 and 1984 labor productivity increased by 35% in the non-electrical machinery sector, by 24% in the electrical machinery sector, and by 11% in the metal products sector compared with only 5% for the industry as a whole. Average salaries increased in a more uniform pattern, around 10 to 11% compared with an industrial average of 9% between 1981 and 1984. In spite of much talk in Cuba about the need to link salary to productivity, the data in Table 6 seem to reject such a claim. Rather it seems that salary increases in Cuban industry are annually decided at a uniform rate, irrespective of sector productivity.

Capital invested in the engineering goods industry is quite considerable. In 1981 the capital invested amounted to 481 million pesos compared with a sector gross value of output of 795 million pesos, suggesting a capital/gross output ratio of 0.60, lower, however, than the industrial average of 0.89 in that same year. By 1984 this gap had widened, implying that both the average and incremental productivity of the capital invested in the engineering goods sector is higher than for industry as a whole. Annual gross investments in the sector averaged 10% of total industry investments during the first three years of the first plan, then increased to around 15% in 1979 and 1980 but the share then gradually went down again to around 10% by the end of the second plan (see Table 3).

The rate of investment has been highest in the non-electrical machinery sector. Capital invested in that subsector increased by 55% between 1981 and 1984 compared with an industrial average of 28%. The non-electrical machinery sector is also the most important capital goods producing sector. According to a special, unpublished survey of the most important enterprises producing capital goods (or *medios de producción* — means of production),[27] no less than 84% of the capital goods produced in Cuba in 1983 originated in that subsector. The survey comprises 45 of the 175 enterprises producing engineering goods in Cuba. The capital goods production of these 45 enterprises accounted for 37% of total engineering goods output in Cuba in 1983. The most important goods produced by the same enterprises in 1983 were: transport equipment (buses, trucks, railway wagons, fishing boats) — 23.8%; agricultural machinery and equipment (combine-harvesters, plows, harrows, trailers, towed scrappers, seed drills and irrigation and planting equipment) — 19.7%; machinery and equipment for the sugar industry (including boilers and mill tandems) — 8.6%. Over half of the output (or 52.1%) was thus accounted for by these three types of machinery, all of which are part of the "non-electrical machinery" branch. Other capital goods covered by the survey included water pumps, industrial tanks, lathes, cranes and prefabricated steel dies and moulds. The capital goods produced by the electrical machinery sector are mainly found in the nascent computer industry which, however, as of 1983 only accounted for some 2% of the total output of capital goods in Cuba.

One interesting aspect of the development of the capital goods industries in Cuba is that an obvious effort has been made to spread the industries around the island instead of concentrating most of the production around the largest city, Havana, as is so often the case in developing countries. Thus, according to the survey, no less than 59% of the capital goods production was outside the city of Havana. As a matter of fact, some of the largest capital goods industries are found in the provinces of Villa Clara and Holguin. The largest capital goods producing plant in Cuba, in terms of employment, is the earlier mentioned Santa Clara mechanical plant (Fábric Aguilar Noriega) with 2,388 people on the payroll, followed by the combine-harvester plant (LX aniversario de la revolución de octubre) in Holguin, in the eastern part of the island, with 2,301 people employed. Most of the larger capital goods producing industries in Cuba were either constructed or largely expanded during the latter part of the 1970s or early 1980s.

4. AGRICULTURAL MACHINERY — THE BASE

It is quite clear that the strategy of building up a capital goods industry in Cuba has been based on the need to supply agriculture with the machinery and equipment, requisite for raising productivity and efficiency. This was envisaged already in 1964 by Ché Guevara when, as Minister of Industry, he inaugurated the Santa Clara Mechanical Plant.[28] The production of agricultural machinery for the needs of not only the Cuban sugar industry, but also non-sugar agriculture, would be the cornerstone in the Cuban capital goods industry, he said. From the earlier mentioned survey it is also clear that machinery and equipment production designated for agriculture account for the lion's share of capital goods output in Cuba. Machinery and equipment for the sugar industry and other agricultural machinery account for almost 30% of production. This, however, is certainly an underestimate of agriculture-related capital goods production, as several other types of machinery and equipment (such as tracks, railway equipment, water pumps, etc.) are also used in agricultural activities.

The role of agricultural machinery as a point of entry into capital goods production is also stressed by UNIDO (United Nations Industrial Development Organization).[29] In a worldwide study on the agricultural machinery industry in 1979, UNIDO classified agricultural machinery into four categories in accordance with the technical complexities involved:[30]

Category A: Hand tools, very simple equipment and animal drawn implements
Category B: Simple equipment and tractor drawn implements
Category C: Machines and items of equipment, tractor drawn machinery
Category D: Self-propelled power machines and stationary equipment of a high technological level

Although this categorization is useful as a guide to the complexities involved in the production of various types of agricultural machinery and equipment, it no doubt oversimplifies the difficulties involved in moving from one category to the other. As Hans Gustafsson has pointed out in an interesting essay criticizing the UNIDO approach,[31] it is not obvious that a country that produces A and B goods automatically will move up to the category C and D levels which are much more technologically complex, requiring different and much more advanced technical skills. A true indigenous capability to manufacture capital goods, such as pumps, valves, turbines, gearboxes, speed-reducer and engines, does not exist unless a country has the technical capabilities to manufacture a significant proportion of the relevant components and subassemblies that together constitute the final product. Consequently, according to Gustafsson,

> in order to serve as a point of entry into the production of capital goods, agricultural mechanization must generate technical capabilities and form an industrial base that is conducive to the manufacture of parts and components which are similar or near similar to the essential constituent parts of other capital goods. It is a necessary condition that

various standard technologies are mastered and that a certain degree of specialization is attained in some activities. This would include, among other things: gear-making, heat treatment, steel foundry, special and precision machining, press forging just to name a few key technologies.[32]

The experience of Cuba is a case in point. Cuba had no doubt already mastered the techniques of producing A and B goods before the Revolution. Then, when the US embargo of the island in the early 1960s made the production of spare parts for existing machinery, especially in the sugar industry, necessary for survival, Cuba "moved up" to the production of C goods. Although it is no doubt true that much of the success in this field can be attributed to "learning by doing," the success would have remained quite limited if Cuba had not at the same time launched a massive educational campaign upgrading the skills of the labor force, among other things, sending thousands of workers and engineers abroad for training.

The real breakthrough came in the 1970s when Cuba started producing advanced machinery and equipment, all of it Cuban designed, for the sugar industry. Already in 1965 a program had been set up in cooperation with several socialist countries for the modernization of the sugar factories. As a result most of the sugar mills were modernized or their capacity expanded in the 1970s. By 1980 obsolete equipment had been reduced to 50% and by 1985 this figure was expected to have decreased to 36%.[33] The year 1976 also marked the initiation of construction of completely new sugar mills, entirely designed by Cuban engineers, and with 60% of the equipment produced in Cuba (most of it at the expanded Santa Clara mechanical plant).

In the 1970s Cuba "moved up" to the production of category D goods (self-propelled power machines), when in 1977 a plant was set up in Holguin for the mass production of sugar combine-harvesters. But this step had been in no way easy. The story behind the success is briefly as follows.

In the early 1960s, just after the Revolution, sugar was entirely cut by manual labor (*braceros*) in Cuba. In order not only to increase efficiency but also to rid the working class of this bondage of "the old capitalist society," the revolutionary government soon made a commitment to rapidly introduce and diffuse mechanization in the sugar fields of the island. The whole decade of the 1960s was, however, filled with disappointments. Much effort went into the design of a Cuban made harvester. The first model, named Henderson after a Cuban engineer, was a simple machine that would be easy to construct in Cuba by adapting a heavy tractor.[34] A total of 148 units were produced for the 1969/70 *zafra* at the Santa Clara mechanical plant. But even with the introduction of these Henderson harvesters, mechanization was as low as 1% during the 1970 *zafra* and since, in addition, the Henderson machine was not very successful (especially since it could not clean the cane from extraneous matter) it was decided to discontinue production in 1972.

However, while working on the Henderson model, the same group of Cuban engineers had been working on the design of a new machine that would be able to cut and clean as well as load efficiently. The test results were so positive that Fidel Castro in 1968 symbolically named the machine *Libertadora* since it would now liberate field labor from the burden of cutting cane by hand. The model was gradually modified and 24 units were produced at the Santa Clara plant for the 1969 *zafra*. The Libertadora was built on the chassis of a Soviet made sugar harvester which had been taken out of operation in 1968. In 1970 more powerful Libertadora models were produced and the 1400 model especially turned out to be quite successful. At about that time negotiations had started with the West German company Claas Maschinenfabrik for a joint collaboration on the manufacture of a modified Libertadora on a large scale. The patent rights were handed over to the Germans in exchange for an offer to allow the Cubans to import the machine at a subsidized price.[35]

Claas Maschinenfabrik was evidently not interested in exploiting the patent rights for the home market (since there is no cane production in Germany). Thus this transfer of technology from the developing country, Cuba, to West Germany meant that Claas Maschinenfabrik could begin mass production of a cane combine harvester for export. Since 1972 Claas has reportedly exported the Libertadora 1400 model to 44 countries, including the United States (which paradoxically — at least officially — still prohibits products of Cuban origin!). Among others 169 units have been sold to Cuba, 121 to Argentina, 99 to the United States (Florida), and 82 to Mexico, just to mention a few markets.[36]

An interesting and controversial question then arises. Why did the Cubans give the patent rights for a machine they had themselves designed to a West German firm? Presumably, there was no other viable alternative open to the Cubans at the time. The Cubans were then simply incapable of anything beyond the basic design stage; they were not able to carry out detailed design, nor were they able to manufacture complicated pieces of equipment on a large scale. The mechanical industry was not developed enough to handle all the sophisticated stages involved. For example, Cuba did not possess an adequate technical capability in the field of hydraulic components which the production of the Libertadora 1400 required. One commentator concludes that "an important lesson of the Libertadora story is that it is easier for a developing country to design a machine indigenously than to produce it on a large scale."[37]

But the story does not end here. In 1969 Cuban engineers had constructed another prototype intended to cut green cane in the field and provide high yields. After some trial runs the prototype was sent to the Soviet Union, within the framework of a collaboration agreement. There two identical prototypes were constructed and after several modifications and further testing in Cuba, serial production of this Cuban–Soviet designed machine, named KTP-1,[38] was begun in the Soviet Union in 1973. In that same year 50 KTP-1 units were sold to Cuba and by the mid-1970s there was a yearly import of some 200–300 units.

The KTP-1 model is a self-propelled machine that can efficiently cut cane at the ground level, chop it into pieces, clean it and subsequently deposit it on a cart which follows the combine. The KTP-1 is similar in principle to the Massey–Ferguson 201, also introduced on a mass scale into Cuba (imported from Australia) in the 1970s.[40]

In 1977 a huge plant was inaugurated by Fidel Castro in Holguin. The plant, named "60th Anniversary of the October Revolution," exclusively produces sugar combine harvesters — at first only the KTP-1 model, but recently also a more sophisticated version, the KTP-2. This plant today produces some 650 units per year, making it by far the largest cane combine harvester plant in the world.

As of 1983, 52% of the cane cutting in Cuba was carried out by combine harvesters (today the figure is around 60%); in 1970 the level of mechanization had been only 1%. Most of the mechanical harvesting is carried out by the KTP-1 (about 86%), the remainder by the MF-201 and the Libertadora 1400 (see Table 7). The total number of cane harvesters operating in Cuba increased from 172 units in 1971, to 1,007 in 1975 (of which there were 422 KTP-1, 418 MF-201 and 167 Libertadora 1400), to 2,423 in 1980 (of which there were 1,901 KTP-1, 365 MF-201 and 157 Libertadora 1400), and to no less then 3,727 units in 1983 (of which there were 3,453 KTP-1, 160 MF-201 and 114 still going strong Libertadora 1400).[41]

5. FROM IMPORT SUBSTITUTION TO EXPORT PROMOTION?

Cuba has undoubtedly made significant progress in building a domestic capital goods industry after the Revolution. It is estimated that the Domestic Procurement Ratio (DPR) — that is, the ratio of apparent consumption produced by domestic industry — for capital goods today is about one-third (see Table 8). Compared with the ratio in 1977 (18.1%), the increase in the DPR has been remarkable, indeed.

It should also be mentioned that there is a slight bias in the estimates since production is recorded in constant (1981) prices while import figures are given in current prices, underestimating apparent consumption (and hence overestimating the DPR) for earlier years.

The Domestic Procurement Ratio is, however, still considerably lower in Cuba than in NICs such as Brazil, Mexico and the Republic of Korea.[42] Cuba will no doubt increase its DPR in the coming years but there is obviously a limit to the possible success of import substitution industrialization in a country with a market of only 10 million inhabitants. A small or medium-sized country industrializing through import substitution must sooner or later begin to export, and capital goods are no exception. There are important economies of scale in capital goods production and a strategy of self-reliance aimed at production for the home market only, might

Table 7. *Mechanization of sugar harvest by type of harvester, 1972–83 (percent)*

	Type of harvester			
	MF-201	Libertadora 1400	KTP-1	Total cut by harvester
1972	4	1	—	7*
1973	1	3	1	11*
1974	10	4	4	18
1975	12	4	9	25
1976	13	5	14	32
1977	13	4	19	36
1978	10	4	24	38
1979	10	4	28	42
1980	9	4	32	45
1981	8	3	36	47
1982	6	3	41	50
1983	5	2	45	52

Source: Information supplied by Ing. Jorge Abreu, Deputy Director of Research of CICMA (Centro de Investigaciones de Construcción de Maquinaria Agrícola).
*Including other types of harvesters (mainly Henderson).

Table 8. *Production, imports and apparent consumption of machinery and equipment** 1963–84 (million pesos)

	Production† (1)	Imports (2)	Apparent consumption (3)	Domestic procurement ratio (4) = (1)/(3)
1963	48.2	110.0	158.2	30.5
1964	49.4	127.7	177.1	27.9
1965	44.0	123.4	167.4	26.3
1966	46.5	123.6	170.1	27.3
1967	49.4	238.7	288.1	17.1
1968	60.8	282.4	343.2	17.7
1969	56.4	360.3	416.7	13.5
1970	84.9	419.0	502.9	16.9
1971	103.2	378.5	481.7	21.4
1972	119.6	216.9	336.5	35.5
1973	161.0	291.8	452.8	35.6
1974	200.1	436.7	636.8	31.4
1975	253.7	852.4	1,106.1	22.9
1976	275.0	964.3	1,239.3	22.0
1977	292.4	1,324.3	1,616.7	18.1
1978	359.2	1,151.7	1,510.9	23.8
1979	420.1	1,133.7	1,553.8	27.0
1980	434.4	1,427.6	1,826.0	23.3
1981	553.4	1,502.0	2,055.4	26.9
1982	642.7	1,447.7	2,090.4	30.7
1983	746.0	1,610.7	2,356.7	31.7
1984	876.2	1,878.1	2,754.3	31.8

Source: Table 1 and *Boletín Estadístico de Cuba* (1968 and 1971); *Anuario* (1974, 1979, 1982, 1983, 1984).
*Non-electrical machinery only (including transport equipment).
†Constant prices.

easily prove both costly and counterproductive in the long run.[43]

But then arises the problem of *how* to succeed on the international market where cut-throat competition usually prevails. On top of everything else, the international market for agricultural machinery, where Cuba is apparently rather strong, is dominated by the large transnational corporations. Cuba will no doubt benefit from its integration with the CMEA countries[44] and thereby find an outlet for exports of some capital goods. This is an important future market for Cuba and was referred to by former Vice-Minister of Planning, Miguel Figueras, in his earlier mentioned study on capital goods production in Cuba.[45] At a recent conference for exporters of the engineering goods industries in Cuba (organized by SIME, the Steel and Mechanical Industry Ministry), it was announced that the Cuban made truck Taino was being tested in the German Democratic Republic for possible future sale to that country.[46]

The most interesting future export market for Cuban capital goods is probably, however, other developing countries, especially the poorer and less developed ones, where Cuba might offer cheaper goods (although not necessarily with better quality) and better terms than the large transnational corporations. As a case in point, Cuba is already selling the KTP–1 harvester to Nicaragua. But would the KTP–1 stand a chance in open competition with, let us say, the Claas-Libertadora? At least price-wise it would, since the cost of production of the KTP–1 (38,000 pesos, or, at the official exchange rate, about US $42,000) is reportedly less than half the cost of production of the Claas-Libertadora.[47]

6. THE PROSPECTS OF CAPITAL GOODS PRODUCTION IN CUBA

As mentioned earlier, capital goods production has accelerated under the last Five-Year Plan in Cuba. Between 1980 and 1985 engineering goods output (ISIC 38) expanded at a rate of

15.6% per year, of which non-electrical machinery grew at a rate of 17.3%, electrical machinery at a rate of 15.0% and metal products at a rate of 10.5%. There is of course, no guarantee that such a high rate of growth will be sustained in the future (although many NICs such as the Republic of Korea, Hong Kong and Taiwan have experienced such high growth rates for 15 years or more), but a projection of this growth trend throughout the decade, and up until the turn of the century, would indicate that the share of engineering goods in total industrial output would increase from 13.2% in 1985 to 18.4% in 1990 and to 36.4% by the year 2000, which would place Cuba at the same level as the industrialized countries are today (see Table 9). Such a projection is, however, of limited interest and can only give a rough idea of the direction of the industrialization process.[48] Many other factors will determine the actual rate of growth of the capital goods industries, such as demand and cost/benefit considerations. As mentioned above, there are important economies of scale in capital goods production and it simply might not be worth the price to produce all important capital goods even if the resource base is available.

There are several important types of machinery and equipment that Cuba does not produce and that are currently imported on a mass scale every year. For instance, Cuba does not produce tractors, bulldozers, or excavators in spite of the fact that such equipment is in high demand in Cuba. Between 1963 and 1984 Cuba imported 150,195 tractors worth 667 million pesos,[49] 9,785 bulldozers worth 296 million pesos, and 8,416 excavators worth 323 million pesos. If Cuba had been able to produce all these machines it would obviously have saved foreign currency. Yet, it is quite possible that the cost of production would have been prohibitive even if the skills and industrial base had existed. An important aspect of reducing costs in not only taking advantage of economies of scale but also making maximum use of economies of scope, that is using one component in several lines of producton. A case in point is the production of diesel engines. The production of diesel engines has begun, although for the time being only on an experimental basis. When full-scale production starts these diesel engines may be used not only for the cane combine harvesters, buses, trucks, and fishing boats which are already produced or assembled in Cuba, but might also make possible the production of tractors, bulldozers and excavators in Cuba in the future.[50]

An important line of production in the future might be machine tools. Cuba does not at the moment produce any sophisticated machine tools although demand for them is quite high, especially in the capital goods industries. Today Cuba has a machine park of some 50,000 machine tools, of which an estimated 30,000 have been imported since 1963, and almost 20,000 since 1974.[51] There is no doubt a large future market for machine tools in Cuba and although it is an industry which requires complex technologies, Cuba is considering manufacturing special machine tools used in the sugar industry — machine tools that might also later by exported.[52] Today most of Cuba's machine tools are imported from the CMEA countries.[53]

Another factor determining the future output of capital goods in Cuba is the availability of a skilled labor force for the production of such goods. As mentioned earlier there is no automatic point of entry into the production of capital goods, even if a long and sustained "learning by doing" stage is quite helpful. For a sustained rate of growth of capital goods production, a skilled and adequately trained labor force is essential. And it is in this respect that the Cuban prospects seem most promising. As seen in Table 10, the

Table 9. *Projected output of total industry and of engineering goods industries 1985–2000 (million pesos, 1981 prices)*

	1985 Value	%	1990 Value	%	2000 Value	%
Total industry	9,997	100.0	15,175	100.0	35,054	100.0
Engineering goods	1,322	13.2	2,790	18.4	12,758	36.4
of which:						
Non-electrical machinery	966	9.7	2,146	14.1	10,583	30.2
Electrical machinery	152	1.5	307	2.0	1,253	3.6
Metal products	204	2.0	337	2.2	9??	?.6

Source: Based on growth rate trend 1980–85 (see Tables 1 and 5).

Table 10. *Educational level of the labor force* — 1982

	With primary education only (6 years) Thousands	%	With lower secondary education (9 years) Thousands	%	With higher secondary education (11–12 years) Thousands	%	With university education Thousands	%	Total labor force Thousands	%
Total labor force	1,046.9	36.4	1,260.5	43.8	399.4	13.9	168.4	5.9	2,875.2	100.0
Industrial labor force	253.6	40.0	296.8	46.8	66.6	10.5	17.7	2.8	634.7	100.0
Labor force in engineering goods industries	26.9	25.4	58.6	55.3	16.1	15.2	4.4	4.1	105.9	100.0
of which:										
Non-electrical machinery	20.8	25.7	45.0	55.5	12.1	14.9	3.1	3.8	81.0	100.0
Electrical machinery	2.2	19.1	6.5	57.3	1.8	16.3	0.8	7.4	11.4	100.0
Metal products	3.9	28.6	7.1	52.6	2.1	15.5	0.4	3.3	13.6	100.0

Sources: Data for total labor force and industrial labor force in *Anuario* (1983). Data for the engineering goods industries: unpublished data supplied by CEE (Comité Estatal de Estadísticas)

level of education of the Cuban labor force is quite impressive considering that Cuba is still a developing country. In 1982, 60% of the industrial labor force had nine years of education or more and in the engineering goods industries this percentage was as high as 75%. Fifteen percent of the labor in the engineering goods industries had completed senior high school (or equivalent technical education), and 4% had a university degree.

The future supply of skilled workers and engineers to the engineering goods industries also looks promising, considering the rapid increase in enrollment in relevant subjects at secondary technical schools and at polytechnical universities (see Table 11). Enrollments have not only increased at large but also in directly capital goods related areas, such as machinery construction and electronics (including automation and communications), subject areas unheard of before the Revolution. Between 1977 and 1984 alone 72,028 students in the field of machinery construction, and 14,136 in electronics graduated from secondary technical schools. In the same period 3,924 engineers specializing in machinery construction, and 3,466 in electronics, graduated from the universities.[54] So if the future of capital goods production in Cuba were to be determined by the future supply of skills alone, the prospects would seem quite bright, indeed.

Table 11. *Enrollment at secondary technical schools and at universities* 1970/71–1984/85

School year	Secondary technical schools Total	Machinery construction	Electronics, automation & communications	Universities Total	Machinery construction	Electronics automation & communications
1970/71	27,566	NA	NA	35,137	NA	NA
1971/72	30,429	NA	NA	36,877	NA	NA
1972/73	41,940	NA	NA	48,735	NA	NA
1973/74	56,959	NA	NA	55,635	NA	NA
1974/75	94,634	NA	NA	68,451	NA	NA
1975/76	114,653	NA	NA	84,750	NA	NA
1976/77	159,440	7,207	3,059	107,091	2,945	2,710
1977/78	194,034	27,243	5,757	122,597	3,376	3,000
1978/79	198,261	25,259	5,464	133,014	3,643	3,164
1979/80	214,615	27,099	4,577	146,240	4,469	3,331
1980/81	228,487	25,466	4,986	151,733	4,821	3,293
1981/82	263,981	29,964	6,131	165,496	4,817	3,372
1982/83	285,765	32,177	6,057	173,403	5,188	3,396
1983/84	312,867	36,106	7,718	192,958	5,588	3,837
1984/85	305,556	34,116	9,330	212,155	6,147	4,186

Source: *Anuario* (1977, 1982, 1983 and 1984).

NOTES

1. Capital goods are often used as a categorization of ISIC group[38] (ISIC = International Standard Industrial Classification of .all Industrial Activities) which comprises "manufacture of fabricated metal products, machinery and equipment" but that is not entirely correct since ISIC 38 also includes durable consumer goods (such as radios, TVs, refrigerators, stoves, etc.). A better label for ISIC 38 is the *engineering goods industries* and this is also the label given to it in this study. Thus the term capital goods is reserved for goods that are used as means of production.

2. UNCTAD (1985), p. 2.

3. Rosenberg (1976), p. 144.

4. UNIDO (1985), p. 1.

5. UNCTAD (1985), p. 24.

6. See Brundenius (1984a) and Brundenius (1986).

7. CEE (1981a).

8. Rodríguez (1980), p. 145.

9. Castro Tato and Bas Fernández (1982), p. 16.

10. *Ibid.*

11. See, for instance, Brundenius (1984b), Chap. 3.

12. Kalecki (1976).

13. Brundenius (1984b), p. 49.

14. For an account of the US economic warfare against Cuba in the 1960s, see Adler-Karlsson (1968), Chap. 17.

15. Information given to the author during visit to the Santa Clara mechanical plant on 18 September 1984.

16. Brundenius (1984b), Table 3.10.

17. Global Social Product (GSP) is the gross value of production of the "productive sphere" (= agriculture, forestry & fishing, manufacturing industry, electricity, construction, transport, communication and trade). GSP thus excludes important services such as education, health and other government services.

18. Figueras (1985), pp. 36–37.

19. See Brundenius (1984b), p. 58.

20. CEE (1981b).

21. *Ibid*.

22. National income (net material product) equals net value of production ("value added") in the productive sphere (see note 17).

23. PCC (1981).

24. Castro Tato and Bas Fernández (1982), pp. 55–56.

25. *Ibid.*, p. 57.

26. See note 1.

27. Survey of capital goods industries prepared for the author by CEE (Comité Estatal de Estadísticas) in September 1984).

28. Quoted in the daily *Trabajadores* (Havana: 3 November 1982).

29. UNIDO (1985).

30. UNIDO (1979).

31. Güstafsson (1986).

32. *Ibid.*, p. 4.

33. MINAZ (no date), p. 20.

34. For a detailed account of the trial and error process of developing a Cuban designed sugar combine harvester, see CICMA (1983) and Edquist (1985).

35. Edquist (1985), pp. 48–49.

36. *Ibid.*, p. 129.

37. *Ibid.*, p. 130.

38. KTP is the Russian abbreviation for Kombain Trostnikouborochny Pryamotochny ("straight forward cane harvesting combine").

39. Edquist (1985), p. 52.

40. *Ibid*.

41. Interview with Ing. Jorge Abreu of CICMA (see Table 7) on 19 September 1984.

42. UNCTAD (1985), Table III.2.

43. UNIDO (1979), for instance, claims that a plant producing tractors should have an annual capacity of 10,000 units in order to be economical.

44. CMEA is the Council of Mutual Economic Assistance (often referred to as COMECON); Cuba has been a member since 1972.

45. Figueras (1985), pp. 53–54.

46. "Las exportaciones en la industria sideromecanica," *Bohemia* 9 May 1986).

47. Interview with Ing. Jorge Abreu (see note 41).

48. A similar projection indicates that the DPR of capital goods would reach 43% in 1990 and 66% in 2000, compared with 32% in 1984 (see Table 8). Likewise it could be projected the Cuban GDP per capita would be 3,045 dollars and 1990 and $6,455 in 2000 (constant 1981 prices), compared with an estimated per capita income of $2,100 in 1985 (based on the assumption that GDP increases at the same rate as National Income between 1980 and 1985).

49. One peso is approximately equivalent to US $1.10 at the official exchange rate.

50. Figueras (1985), pp. 102–103.

51. The machine park figure is given by Figueras (1985, p. 39). The imports of machine tools figures were calculated by the author from *Boletín Estadistizo de Cuba* (1968, 1970, 1971) and *Anuario Estadistizo de Cuba* (1974, 1979, 1981 and 1984).

52. Figueras (1985), p. 104.

53. *Anuario* (1984), p. 365.

54. See Brundenius (1986), Appendix Table 5.

REFERENCES

Adler-Karlsson, Gunnar, *Western Economic Warfare, 1947–1967* (Stockholm: Almqvist & Wiksell, 1968).

Brundenius, Claes, "Capital goods in economic thought — Some notes from selected readings," Discussion Paper Series No 164 (Lund, Sweden: Research Policy Institute, University of Lund, December 1984a).

Brundenius, Claes, *Revolutionary Cuba — The Challenge of Economic Growth with Equity* (Boulder: Westview Press, 1984b).

Brundenius, Claes, "The role of capital goods production in the economic development of Cuba," *Political Power and Social Theory: A Research Annual*, Vol. 6 (Greenwich, CT: JAI Press, 1986).

Castro Tato, Manuel, and Arturo Bas Fenrnández, "The development of the capital goods industry in the Republic of Cuba" (Vienna: UNIDO, 2 November 1982).

Centro de Investigaciónes de Construcción de Maquinaria Agricola (CICMA), *Desarrollo perspectivo de las combinadas cañeras y sus limitaciones* (Havana: 1983).

Comité Estatal de Estadisticas (CEE), *Anuario Estadistico de Cuba* (Havana: 1974, 1979, 1981, 1982, 1983, 1984).

CEE, *Boletin Estadistico de Cuba* (Havana: 1968, 1970, 1971).

CEE, *Algunas concepciones sobre el desarrollo de Cuba en la década de 1950 — Recopilación de textes, Tomos 1–4* (Havana: 1981a).

CEE, *Cuba — Desarrollo economico y social durante el periodo 1958–80* (Havana: 1981b).

CEE, *Cuba en cifras* (Havana: 1981c).

CEE, *Boletin Estadistico Mensual* (Havana: noviembre de 1985).

Edquist, Charles, *Capitalism, Socialism and Technology — A Comparative Study of Cuba and Jamaica* (London: Zed Books Ltd., 1985).

Figueras, Miguel Alejandro, *Producción de maquinarias y equipos en Cuba* (Havana: Editorial Cientfico-Técnica, 1985).

Gustafsson, Hans, "Is agricultural mechanization a point of entry into capital goods production in LDCs? — A critique of the UNIDO approach," Forthcoming Discussion Paper (Lund, Sweden: Research Policy Institute, University of Lund, 1986).

Kalecki, Michal, "Hypothetical outline of the Five Year Plan (1961–65) for the Cuban economy," in Michal Kalecki, *Essays on Developing Economics* (London: Hassocks, 1976).

Ministerio de la Industria Azucarera (MINAZ), *Development of the Cuban Sugar Industry* (Havana: n.d.).

Partido Comunista de Cuba (PCC), *Socio-economic Guidelines for the 1976–80 Period* (Havana: 1976).

PCC, *Socio-economic Guidelines for the 1981–85 Period* (Havana: 1981).

PCC, *Lineamientos economicos y sociales para el quinquenio 1986–90* (Havana: 1986)

Rodríguez Mesa, Gonzalo, *El proceso de industrialización de la economía cubana* (Havana: Editorial de Ciencias Sociales, 1980).

Rosenberg, Nathan, *Perspectives on Technology* (Cambridge: Cambridge University Press, 1976).

·7·

The Cuban Health Care System: Responsiveness to Changing Needs and Demands

SARAH M. SANTANA

1. INTRODUCTION

For the last two decades Cuban medicine has been an example of the successes that can be achieved in an underdeveloped country. The improvements in the health status of the Cuban people in the last 25 years have been extensively described elsewhere.[1]

As the economy developed over the last 27 years and living conditions of the population changed, the people's health profile also changed. The corresponding transformations of the structure of the health care delivery system constitute a good case study in responsiveness to population pressures and changing needs. In an unorthodox and idiosyncratic manner the Cuban health system seems to be endowed with the ability to evolve in order to solve problems, eliciting enthusiastic population participation and approval. This plasticity in the face of a dynamic situation, a measure of the success of any system, has caused medical care in Cuba to develop in phases dissimilar from what many advocate as the best road for the Third World.

The basic principles of the system and its underlying assumptions have not changed since 1960. What has changed has been the organizational forms in which those principles are embodied. The most important and urgent needs at any particular moment determined the strategy of health delivery followed.

Cuba's health system is based on five main principles:
1. Health care is the right of the people, thus access is universally equal and free of charge.
2. Health care is the responsibility of the state.
3. Preventive and curative services are integrated.
4. The population participates in the development and functioning of the health care system.
5. Health care activities are integrated with economic and social development.[2]

These principles are still implicit in the health care system. Improving and changing conditions (material, educational, technological, etc.) have allowed for the implementation of the practical measures that developed and refined the tenets of the system.

Recent dramatic changes have taken place at both ends of the medical care spectrum: at the level of primary care and at the level of tertiary care and basic research. After many years of using primary care teams consisting of internists, obstetrician-gynecologists, pediatricians and other specialists, Cuba has finally adopted the use of family practitioners. Simultaneously, great attention and resources are being devoted to sophisticated, expensive and technologically-intense treatment and diagnostic modalities (such as organ transplants, the use of genetic engineering products, *in vitro* fertilization) and to the basic research necessary to sustain these activities.

This paper examines some of the factors that have led to and made possible this latest reoganization of health services delivery. The ways in which these changes have taken place clarify the workings of the system. An analysis of the evolution of the Cuban health system helps to define important issues in the development of health care services of other developing countries.

2. STAGES IN THE DEVELOPMENT OF THE CUBAN HEALTH CARE SYSTEM

(a) *1959–69*

In this initial period the tasks at hand seemed almost obvious: to deliver as much care as possible (mostly curative) to a population suffering from acute infectious diseases; to develop a coherent network of services that would become the infrastructure of a comprehensive, unified health care system; and, considering the need for physicians and other health workers (exacerbated by the exodus of approximately 3,000 physicians in the early 1960s), to train personnel in large numbers as quickly as possible.[3]

In order to solve the high mortality and morbidity problems, acute and episodic care was emphasized. Resources were devoted to maternal–child health and control of infectious diseases through national disease-specific programs and campaigns coordinated with services at the local level. These activities were targeted at particular health problems and specified underserved areas. The process of consolidation of the health care system covered this period. Different types of existing health services were gradually integrated into the services provided by the Ministry of Public Health (MINSAP), including mutualist clinics (organizations providing pre-paid services), and separate state public health programs (such as those for tuberculosis, leprosy, and maternal–child services).[4]

(b) *1970–74*

Health resources in Cuba were finally consolidated into a single system in the late 1960s, marking the beginning of a second phase in the development of health services in Cuba. This second period was characterized by more coherent, better organized services, new programs in maternal–child health, special emphasis on development of primary care centers (policlinics) and better informational and statistical control. Virtually complete vital registration was achieved. During this period mortality and morbidity rates steadily declined. Life expectancy increased, mainly determined by falling mortality rates among children and from infectious diseases (see Table 1). The system was still, however, hospital centered, without sufficient integration of preventive and curative services, and with uneven, incomplete regionalization.

(c) *1975–84*

In 1975 important changes in the Cuban economy and the administration and organization of the government took place.[5] Locally elected governing bodies were created which in turn elected provincial and national legislative assemblies in pyramidal fashion ("*Poder Popular*"). The health system also underwent significant reorganization during this period. This new stage in the development of health services, called Medicine-in-the-Community, was specifically designed to achieve comprehensive delivery, universal population coverage and to complete the regionalization process. Its main features were administrative decentralization and a restructuring of the policlinic, now carrying the responsibility for the total health services of a specific area and population, as well as for expanded teaching and research activities. Cooperation with mass organizations and accountability to local government were improved.

Primary care services were delivered by four specialists: internists, pediatricians, obstetrician-gynecologists, and dentists. These specialists, plus nurses, psychologists and other professionals made up the team responsible for individual health care. Family practice was not implemented, in part because of the difficulty in integrating this physician within secondary and tertiary care structures. Traditional medical belief held that specialists provided better care, and the population expected to receive care from specialists, since this had been the prevailing pattern before 1959. Even those who could not afford a physician were aware that specialty care was the desired standard.

The model of Medicine-in-the-Community established guidelines for primary care specialties:

1. Preventive and curative services were to be integrated.
2. The policlinic was given responsibility for the total health of a defined geographical area and population.

Table 1. *Selected indicators of the health status of the Cuban population*

A. Percentage population over 65 years of age:

1960	1970	1980	1985
4.8	5.9	7.3	8.0

B. Mortality rates per 1,000 population by age:

	1970	1985
under 1*	38.7	16.5
1–4	1.3	1.0
5–14	0.5	0.5
15–49	1.0	1.2
50–64	9.3	9.4
65+	52.9	49.6
Total	6.6	6.4

C. Percentage of deaths over 50 years of age:

1960	1970	1980	1985
60.8	65.9	76.7	78.2

D. Mortality rates per 100,000 population from 10 main causes of death:

	1970	1985
Heart disease	148.6	189.8
Malignant tumors	98.9	117.0
Cerebrovascular disease	60.3	62.3
Influenza and pneumonia	42.1	45.4
Accidents	36.1	42.4
Suicide and self-inflicted wounds	11.8	21.7
Diabetes	9.9	15.3
Perinatal conditions	41.7	12.7
Congenital anomalies	14.1	8.8
Asthma, bronchitis and emphysema	12.5	8.2

E. Mortality rates per 100,000 population for other selected causes of death:

	1962	1970	1985
Direct maternal mortality	118.2 (1960)	70.4	31.3
Acute diarrheal diseases	57.3	17.7	3.0
Infectious and parasitic diseases	94.4	45.4	11.6

F. Incidence per 100,000 population of selected reportable diseases:

	1965	1985
Typhoid fever	3.0	0.6
Tuberculosis	63.5	6.7
Leprosy	4.2	3.5
Diphtheria	8.0	0.0

continued

Table 1. (*continued*)

Whooping cough	26.6	1.9
Tetanus	6.5	0.1
Tetanus neonatorum	1.3	0.0
Measles	118.8	28.6
Malaria	1.6	—†
Meningococcal meningitis	0.3	8.0
Gonorrhea	8.9	361.0
Syphillis	29.7	62.8
Poliomyelitis‡	—	—

G. Life expectancy:

1960–65	1970–75	1980–85
65.10	70.93	73.59

H. Utilization of health services:

	1970	1975	1980	1985
Average yearly visits per child under 1	3.0	6.1	8.3	9.8
Average ambulatory visits per person (all ages)	2.5	2.7	3.1	3.9
Average obstetric visits per delivery	7.0	9.5	11.4	12.8

Source: MINSAP *Annual Reports* (1980, 1985).
*Per 1,000 live births.
†There have been no domestic cases for many years, only a few imported ones.
‡Polio has been eradicated in Cuba since annual immunization campaigns were begun in 1962.

3. Primary care services were to be coordinated with secondary and tertiary levels of care.
4. Continuity of services by the same staff was to be assured.
5. Strict follow-up of high-risk patients and the chronically ill was required.
6. Specialists would practice within the framework of the health team.
7. Services had to include active community participation, especially in health education.

Obviously, these guidelines were the logical extension of the basic principles of the system.

3. THE HEALTH CARE SITUATION IN 1984

By 1984 Medicine-in-the-Community had been in place for 10 years. Regionalization and integration of the health system with other sectors of society had been achieved. The development of the health system, hand in hand with increased decentralization and development of local government ("*Poder Popular*") facilitated better understanding of local needs at the national level and better local response to problems. Undoubtedly, diverse activities, such as preventive and curative services, health education, occupational and environmental health programs and research were better developed and coordinated at the policlinic level. Health educators, psychologists, sociologists, etc. had a place in primary care practice.

The population had become accustomed to new levels of participation in health care. Goals and objectives were those of a model system, thus raising people's expectations. Cubans used (and in some cases, overused) their health system. Patients sought care from the provider of their choice, unrestricted by geographic area. The country's health resources were freely at the disposal of the population and encounters per person continued to rise yearly.

The policlinic, however, did not fulfill one of

the main pre-requisites for effective community-oriented primary care: that it be small enough to facilitate the integration of its staff within the community it serves.[7] One policlinic could serve as many as 30,000 while a team was responsible for a "sector" including 4,000–5,000 people (1,000 children, 2,000 persons for adult care and 2,000 women for ob.gyn.).[8] Thus, many of the services offered fell short of expectations.

There was in fact little continuity of care, a function of the fragmentation resulting from the use of several specialists in the team without a case coordinator or principal provider. The perception of the population (accurate to some extent) that the hospital still drew the best specialists and provided the quickest and best curative care contributed to the inappropriate use of emergency rooms as outpatient clinics, a common enough phenomenon in most countries. Patients were lost to the policlinic after referral to other institutions for specialty care.

Patients initiated encounters by going to the policlinic or to the emergency room for acute, curative care when they were ill. It was a rare well person who used individual services in a preventive manner. Individual health maintenance was neither a habit of the population nor a feasible activity for the policlinic, given its organizational structure and responsibilities.[9]

In order for preventive services to reach full coverage — not possible through individual encounters — immunization and screening campaigns such as those for cervical cancer were necessary. These collective preventive services had always been well executed, used and supported.

The health team did not in fact know the *population* of its sector very well. It was relatively passive in its contacts with patients, although, infants, pregnant women and certain chronically ill patients like diabetics were aggressively followed.

As a result, morbidity surveillance was deficient, causing certain gaps in the available information regarding the health status of *populations* (it is necessary to emphasize that information on *patients* had been abundant and accurate for a number of years). In the absence of nationwide surveys some silent conditions such as hypertension, or risk factors such as obesity, were seriously underreported or unknown. This hampered planning efforts and impeded good community diagnosis.

The health profile of the population had changed. The main causes and age distribution of deaths as well as the morbidity profile shifted from those classically associated with underdevelopment (infectious disease, high percentage of deaths under five years of age, high maternal and infant mortality, etc.) to those typical of a developed society. The population now experiences chronic diseases that kill at a later age, that have long, silent latency periods and require early, carefully planned and long-term prevention strategies. These are illnesses less amenable to quick, low-technology cures.

Degenerative diseases caused by so-called lifestyle risk factors, are a response to the stresses of modern life, diet, etc. Their ultimate control lies in long-term preventive programs that should take into account damaging occupational, environmental, and behavioral exposures as well as protective factors. These programs are very different in design from the relatively short-term programs (like immunizations) used to control many infectious diseases. The policlinic was not well equipped to meet this new challenge.

Because the problems were systemic and organizational, the solution had to be national and not local. In the same way that with each stage of development the system had modified its structures to adjust to new necessities, it now had to evolve into the sort of organization that could provide the patient with a practitioner responsible for care, while also penetrating the intricate social web of the community to provide it with collective services.

Not all were new problems. Medicine-in-the-Community had tried to grapple with some of them. The new structures would have to provide solutions to the outstanding issues:
1. The changing health status of the population.
2. The effective inclusion into curative and preventive care of the family, socio-economic factors, the community and occupational and environmental conditions (a responsibility the health team had not been able to discharge adequately).
3. The inability of the policlinic to provide continuity of care, health maintenance and morbidity surveillance.
4. The excessive pressure on and inappropriate use of emergency rooms.
5. The necessity of a smaller-scale entry point to the system, facilitating a more intimate relationship between practitioner and patient and practitioner and community.

4. 1985 — THE FAMILY PHYSICIAN PROGRAM

The institution of the family physician program in 1985 has been as profound a change as the health system has ever experienced. It has meant changing the way primary care is delivered, roughly doubling the number of practicing physicians, increasing nursing staff, changing the role of the policlinic and readjusting administrative, informational, educational and supervisory procedures throughout the entire system.

Cuba presently has 23,000 physicians, or an average of one per 435 persons. The distribution of these physicians is relatively uniform throughout the country, excluding those engaged in research or teaching activities, and about 3,000 doing internationalist work in 26 or 27 foreign countries.[10] By the year 2000 Cubans expect to have 20,000 family physicians practicing at the neighborhood level.[11] This means literally a physician in almost every city block — Cuba is over 70% urban[12] — and easier accessibility to the rural population, for whom the physician-to-population ratio will be even larger. Schools and factories and other work places will also have a doctor on the premises.

Expected to cover the entire country by the year 2000, the family physician program began with two pilot projects, one urban and the other rural. The present program is the result of modifications and adjustments effected over the last two years in those and other additional projects. Each family physician (who will eventually graduate from a residency program in family medicine or "*medicina general integral*") practices together with a nurse in the community in a small office-consultation room. They will each have living quarters nearby or in the same building. Each team is responsible for the preventive, curative and environmental health of a city block or rural area which covers a population of 600–700 persons or 120 families.

Based on an exact census of their populations, family physicians do continuous community diagnosis of their area, collecting health and socioeconomic information for every person, sick or healthy. For the first time, and in an ongoing way, the well population has become part of the health worker's mandate. This is important, especially if health is to be defined as a state of well-being and not merely the absence of disease. It gives health personnel a tangible and real sense of denominators.

Once the system is established throughout the country, data can be available with uniform definitions and standardized collection procedures to be aggregated at different levels as needed. The databases generated by such a system could be an epidemiologist's dream, providing information about diseases and risk factors on a population basis. Such information can yield important insights into the development of disease and the maintenance of good health. Prevention, intervention and research strategies at the local, provincial and national levels will be greatly facilitated, since complete and efficient morbidity surveillance will be possible.

Family physicians have office hours every morning, one night a week and one Sunday a month. Every afternoon they make field visits, accompany patients to specialty consultations and follow patients in hospital. The practitioner constantly addresses the anticipated needs of the healthy population through counselling, health education and other preventive strategies *before* illnesses (or such conditions as pregnancy) arise. They initiate, monitor and participate in educational, environmental, and preventive programs such as exercise programs, activities for adolescents, elderly groups, etc. They work closely with the Federation of Cuban Women, the Committees for the Defense of the Revolution (block associations of residents), and other mass organizations and carry out regular health inspections of community facilities like food stores, pharmacies and day care centers in remote areas.

In areas where the program has been established, the policlinic's direct primary care activities have diminished in favor of those of a back-up facility, providing specialty consultations, teaching, and laboratory and other ancillary services. The policlinic is the support institution for the family physician, playing a strong teaching role in the physician's residency training. Total encounters have increased although the numbers actually taking place at the policlinic have decreased. Laboratory and radiological use per patient visit have decreased and management of patients using multiple medicines has improved as a result of better history taking, longer time with the patient and a better patient–physician relationship. Some hospitalizations have been avoided by keeping the patient at home under the close supervision and frequent (at least once daily) visits of the nurse and/or physician.[13]

The bulk of the cost of the program comes from education, construction, equipment and supplies, not labor. Physicians' salaries in Cuba are modest in comparison with those in other countries.[14] Although some savings may be realized in prevented hospitalizations and lower costs per encounter (there is no pressure on physicians to cut costs), it is likely that total health costs will rise as a result of increasing utilization.[15] Cubans emphasize, however, that the reasons for the establishment of the program are not savings, but better health and more humane medical services. The savings will accrue to the society in the long run in terms of lower mortality and morbidity.

5. DEVELOPMENTS IN TERTIARY CARE

A second aspect of the changes which are taking place in Cuban medicine involves increased emphasis on tertiary care. As in primary care, this "revolution" in medically-related scientific activities has been a response to the health needs of the country. In the same manner that the changing illness profile demanded new primary care strategies, it also required new secondary and tertiary care treatment: effective heart disease and cancer therapies are expensive and require sophisticated, highly specialized research facilities and personnel. Without the necessary curative resources and facilities the primary care structures would fall apart at the point of referral of the patient to higher levels of care.

Resources have been increased and reallocated to support such activities as basic biological research, genetic engineering, specialized training in new diagnostic and treatment techniques and supportive research activities (among them organ transplants, microsurgery and computerized axial tomography, development of instruments and automated equipment, production of biological substances, breeding of animals for research purposes, etc.).[16] Research plans among medical schools and other research institutions have been coordinated and adjusted accordingly.

It is not only a response to internal population needs that has prompted the emphasis on tertiary care and research. Cuba has explicitly stated its goal of developing into a "medical power" capable of. providing the expertise and technological transfer that Latin America and other developing countries may need. Both primary and tertiary care changes help in this purpose.

6. POPULAR PARTICIPATION AND ACCEPTANCE OF THE NEW CHANGES

The population has eagerly welcomed and participated in the new programs. As consumers they have been the recipients of great benefits: access is easier, they are visited at home when ill (not only for preventive services), and the patient has an accompanying advocate at the time of hospitalization or specialty consultations. The closer patient–physician relationship encourages better patient compliance with treatment.

Population acceptance of the new physician is such that even when patients venture on their own to the hospital in search of care, they return to the block to "consult" the family physician before beginning the treatment prescribed by the hospital specialist.

Besides actively participating in the program as patients and consumers, the Cuban people are building the office and living quarters of the physician and nurse with volunteer labor in their spare time. Some of these building brigades are made up of retired persons who feel they want to contribute in some way to the success of the program. This is the most objective and concrete proof of their glad acceptance of the program.

It is not only in the primary care program that the population is participating. At the tertiary care level, new research institutes and medical school additions are being built with the help of volunteer labor. Parents and children are participating in the new college-preparatory high schools organized to train students in the sciences as part of the emphasis on basic research. Workers from many sectors (construction, foreign commerce, agriculture, etc.) are making special efforts in order to meet the needs of the new schools and research facilities.[17]

7. NECESSARY CONDITIONS FOR RESTRUCTURING PRIMARY CARE IN 1986

Why were some of these changes not carried out earlier? It has been suggested that the Cuban health system should have begun using family practitioners years ago. The family physician program could not have been implemented earlier for some very important developmental reasons regarding the availability of material and human resources.

A minimum number of physicians is necessary in order to provide adequate coverage and specialist back-up for the program if it is to preserve the principles of equal and complete access. The educational resources to graduate sufficient numbers of well-trained physicians were not fully in place until recently. The program would have suffered from the same ills as the policlinic if there had not been enough health workers to keep the physician-to-population ratio small.

In addition, the *type* of physician available today is very different than in previous periods. There is an implicit ideological orientation in family medicine, and especially so in a socialist country like Cuba. The family practitioner sees health behavior and illness in a societal context. There must be a strong commitment to communication with the patient, since that is the physician's principal tool in evaluating the psychosocial influences on the person's health. The practitioner must understand the larger systems of family, collective and community in which the individual functions. In other words, the practice of medicine in this context must be carried out by someone convinced of the importance and usefulness of sociological and psychological skills in managing health and disease.

The family physician must be accepted into an already organized community with its networks, formal and informal organizational structures. For this to happen, the physician must "blend in." In order not to risk the imposition of medical and class views upon non-medical issues in the community, the physician must not be a special member of society (however separated from the rest by training), but another worker, from the same class extraction as the rest of the community. It is now that a generation of physicians has emerged which was born, raised and educated within the Revolution and relatively free of the socializing trappings of capitalist medical education. A majority of today's physicians are children of peasants and workers, not the traditional children and grandchildren of doctors.

There is still another characteristic of this generation of physicians that, although not necessary, contributes greatly to the success of the program at this time: the high percentage of women family physicians. Much of the work of the family practitioner includes tasks and responsibilities traditionally considered female: the support of family and community networks, preventive and nursing care, supportive care of the young and elderly, general health education, nutritive and sanitary care of the family, vigilance of the growth and development of children, etc.

The role of women in Cuban society has changed (see Table 2). The majority of family physicians and medical students are women. Their high-quality performance without the loss of humanist values (often erroneously considered "feminine") has encouraged and facilitated the establishment of family practice.[18]

Especially in the more rural and less developed areas, it sometimes may be easier for female physicians to be accepted into a community and into homes where networks are still largely maintained by women. In these areas the traditional healers, midwives, etc. had been mostly women, and although they no longer exist as such their role may not have disappeared from the community's memory.

On the other hand, the number of male nurses

Table 2. *Selected statistics on the participation of women in Cuban society*

1. Women as a percent of the labor force:
   1970: 18%*   1985: 37%

2. Women in the national leadership of important organizations:

	1974	1984
CTC (Cuban workers labor union)†	7.0%	17.7% (1983)
ANAP (National Association of Small Farmers)	2.0%	11.0%
CDR (Committees for the Defense of the Revolution)‡	19.0%	31.8%
UJC (Young Communists League)	10.0%	27.1%
PCC (Cuban Communist Party)	5.5%	12.8%

3. Women in the National Assembly have remained at approximately the same percentage in the last 10 years:

   1976 — 21.8%
   1984 — 22.8%

Source: Unless otherwise noted, Espin Guillois (1986).
*Source: Juceplan (1975).
†45% of shop stewards are women.
‡These are block associations which have responsibilities for tasks such as environmental sanitation, public health campaigns like immunizations, screenings and blood donations, recycling of useable materials, nightwatch patrols and surveillance and political study groups. They are the basic unit of community organization in Cuba and have political and non-political functions.

has increased, a sign of the growing homogeneity of roles in the society. Male family practitioners, like male nurses or teachers, have an additional part to play as role models of men working in professions with nurturing tasks and humanist values. Finally, and perhaps most important, there had to be an organizational back-up structure in place to successfully implement the family practice program. The development of Medicine-in-the-Community laid the groundwork for the family physician's community work.

The health care structure is now an intricate, regionalized network that allows for supervision, professional support, continuing education, vacation and illness coverage for the physician and provides specialty care and hospitalization for the patient. An excellent statistical reporting system was already in existence, facilitating the standardization and integration of new data generated by the program. Now the family practitioner becomes the frontline health worker, but not devoid of the support of an adequate structure. S/he is only the first visible tip of a well-organized and supportive health care "iceberg."

8. RESPONSE AND ACCOUNTABILITY OF THE SYSTEM TO THE POPULATION

How have these changes been a response to population pressure and demands? The family physician program was administratively initiated at the highest decision-making levels in the country, receiving the direct attention of the President and the Minister of Public Health. It was a response to unrest and dissatisfaction among the population expressed through their behavior and by direct complaints to local and national authorities, through *ad hoc* and officially established channels. The government responded.

The people also exerted pressure in favor of highly developed tertiary care. The expectations of treatment modalities and diagnostic procedures such as sonograms, prenatal screenings, and CAT "scans" have been encouraged by the system itself. They are a direct result of the regular and explicit reaffirmation by the leadership of the commitment to free and complete access to the best treatment. The country's aspiration to become a "medical power" also nurtures the expectations of the population.

Thus, the Cuban health care system changes from the bottom up and from the top down. The middle layers sometimes offer varying degrees of resistance to change because they are more insulated than the frontline workers from the demands and needs of the population. They may lack the top leadership's overall long-range vision and knowledge of resources and general country conditions, but they finally yield in the face of the overwhelming logic and momentum of the new measures.

It is precisely this feature — that change comes to the system simultaneously from two directions — that is one of the most effective and refreshing characteristics of the Cuban system. This is dependent on open and effective channels of communication between the population and the leadership of the country.

These channels are structured and unstructured, governmental and non-governmental. They provide multiple paths by which the same problems are vented and brought to the attention of the leadership from different points of view.

Structured governmental mechanisms include those of local elected governments (*Poder Popular*). Each delegate meets every six months in general assembly with all the population of his or her district (in addition to the time devoted to the problems brought to his/her attention at the *Poder Popular* office weekly). The problems that cannot be solved at the local level are then channeled to higher governmental levels. Health areas and delegate's districts coincide so that there is accountability in health services delivery to the elected body and to the population.

Each workplace is accountable through administrative structures to local, provincial or national government levels (the state is the principal employer in Cuba) and the concerns of each workplace — from the administration's point of view — are channeled that way.

The role of the Cuban Communist Party (and the Union of Communist Youth) in Cuban society goes beyond the scope of this work. Although it is a non-governmental organization, it sets policy. Suffice it to say that it provides channels through which the perceptions of the membership are funneled directly into the framework of the highest policy-making bodies.

Non-governmental structures include the union sections at each workplace, the block associations, the federation of women, the association of small farmers, children and students organizations, etc. These organizations are independent conduits of the concerns of their membership, and each organization has mandated representation at the different levels of the government and the Party.

Unstructured, spontaneous mechanisms of communication up and down the societal structures are many and are equally initiated by the population and by the leadership. They include the advocacy of special interest associations, like the association of inventors and innovators, or the association of blind persons, or professional groups. Letters, calls, telegrams are sent to the press, to ministers and other officials, to Party organizations, to the Council of State and Ministers. The leadership constantly visits workplaces without advance notice to gather firsthand knowledge of problems. Impromptu meetings and large assemblies happen often in which specific sectors of society such as cooperative farmers, construction workers or hospital staff discuss problems and policies with government officials at all levels.[19]

Although some of these channels seem to defy institutionalization, they work efficiently and become traditional methods of communication within any society. They are akin to the constant correspondence between legislators and their constituents, to the lobbying by such broad-based organizations as labor unions, business groups or those grouping retired persons. In any society these channels of communications provide a stronger and more continuous dialogue between the population and its government than periodic elections every two, four or six years. In Cuba they exert very powerful influence on governmental behavior and policy. Because of the massive nature of the membership of many of these organizations and the high level of participation by most of the members, the very large majority of the population is represented through one or another group.

Health is not the only example of this mechanism at work. Measures legislatively or administratively enacted by the Cuban government, such as housing and educational reform, regulations upon retail sales and laws defining the relationship between workers and management, have had their genesis and have been periodically adjusted through these democratic mechanisms. The state is thus able to respond quickly and effectively to population needs and the population participates in actions taken as direct consequence of its demands.

9. CONCLUSION

Unified, comprehensive care, for which one single health provider is responsible (whether physician or not) is not a new concept. Its genesis goes back to the 19th century.[20] It has been advocated as a method to deliver inexpensive care in developing countries if non-physician personnel are used. Such programs have been deemed effective with providers known variously as *"feldshers,"* *"promotoras de salud"* and "barefoot doctors."

In developed countries the family physician has been a growing phenomenon, together with nurse practitioners and physician assistants. They all have a wholistic concept of the patient, which includes the family and the wider environment.

However, we now face a different practitioner. The Cuban family physician has responsibility for a geographically defined population, not just patients. S/he has the complete support of an intricate network of occupational, environmental and personal health services. The work is integrated with non-medical sectors of the community and prevention activities are of paramount importance. Eventually the whole country will be covered in a coordinated manner.

This program, though it exists in a Third World country, goes further than programs even in developed nations. It defies some of the current thinking about the organization of health care services in poor countries. In Cuba a primary care general practitioner was not put in place until the structure of the health system had developed. Specific primary care programs were developed by disease and specialty first. Twenty-five years ago Cuban health workers were carrying out programs similar to those advocated today by UNICEF (for example, anti-diarrheal and immunization campaigns). However, Cuba was implementing these vertically-administered programs while simultaneously building the structure of a budding health system, the ultimate and explicit goal of which was the provision of comprehensive care. Thus, the vertical programs eventually became an integral part of regular services.

Categorical disease-specific or area-specific programs can be an effective way to tackle short and long-term medical problems in developing countries *if* there is concurrent development of a system with clear principles and goals to eventually provide complete health care to the population. Otherwise, they fall short of expectations, for lack of health-supporting structures.

All this, of course, is only possible if there is a political will to provide services to the maximum degree possible, even at the cost of some alternative private or public economic activities.[21] Cuba has implemented its social and economic development programs (health services, education, construction of the physical infrastructure, industrial and agricultural facilities, etc.) at the cost of postponing other types of expenditures, for example, the importation of luxury goods, or individual housing construction.

The family physician program represents a large investment in human resources. It is not, however, a way to employ an excess of physicians (the so-called "doctor glut"). In Cuba the number of medical graduates is calibrated to fulfill the needs of the country. It is Cuba's long-term goal to provide all professionals with a sabbatical year for study and rest. Today, all state workers recieve 30 days' paid vacation. It is necessary to have a labor force in reserve to cover such absences. Some family physicians will, after a few years, go on to specialize in other fields such as cardiology or immunology. The country supplies physicians to other nations, wherever needed and requested. Under such conditions, great numbers are needed.

The Cuban health system and the family physician program will go through further adjustments: perhaps the size of the population under the care of a nurse–physician team may vary, according to age structure, community diagnosis and population density; the education of the new physician will have to be adapted to the experience drawn from the program; the system for data collection may be modified if necessary. But it is clear that we are before a program with the potential to provide top quality health care in a developing country. It can do so in direct response to population needs and demands, at a relatively modest cost, with virtually universal coverage and a high degree of popular participation.

NOTES

1. See Table 1, Aldereguia (1983), Ubell (1983), Danielson (1979), Stark (1978), Stein and Susse (1972), Navarro (1972), and Hollerbach (1983).

2. IDS/DIR (1977) and MINSAP (1969; 1970).

3. Direccion Nacional de Estadisticas, MINSAP. Datos del Registro de Profesionales.

4. Capote Mir (1979) and Lopez Valdes (1966).

5. DOR (1975).

6. MINSAP (1977).

7. Kark (1981), p. 13. Chapter 1 is a succinct discussion of community-oriented primary care and the necessary conditions for its practice.

8. Gilpin and Rodriguez (1978).

9. The shortcomings of the system and the inception of the family physician program were the subject of panel discussions at the II National Congress of Hygiene and Epidemiology, in Havana (see *Granma*, 16 and 17 October 1985) and the International Seminar on Primary Health Care, also in Havana (see *Granma*, 13 June 1986).

10. MINSAP (1985), and Grundy (1980).

11. Castro Ruz, F. Speech at the II National Congress of Hygiene and Epidemiology (*Granma*, 16 and 17 October, 1985).

12. MINSAP (1985).

13. Rodriguez Abrines (1985).

14. *Gaceta Oficial de Cuba* (1980).

15. Health expenditures per person in Cuba are approximately 60 pesos per year. Gutierrez Muniz (1984) and Castro panel discussion at the International Seminar on Primary Health Care (9 June, 1986). In 1982 6% of the state budget and 2.6% of the global social product was devoted to health (Gutierrez Muniz, 1984). This is not an inordinate amount (see Cochrane, 1983 and Grines, 1983).

Since the beginning of the present health system in 1959 Cuba has kept costs down by the use of volunteer labor through popular participation; the domestic manufacture of equipment, supplies and pharmaceuticals; modest salaries; minimization of duplication of services; targeting of resources to neediest areas and most acute problems (Salimano *et al.*, 1984). In the initial urban pilot project of the family physician program, costs per encounter decreased from $3.84 to $2.94 pesos. Hospital costs in 1985 were approximately $30 pesos per day per patient (Rodriguez Abrines, 1985).

16. and 17. *Granma* (2, 3, 26, 27 July 1986). Speeches by Castro and others at the inauguration of the Institute of Biotechnology and Genetic Engineering (1 July 1986) and at the July 26 celebration.

18. See Table 2. Also, Castro speech and discussion at the International Seminar on Primary Health Care (*Granma*, 13 June 1986). For a description of the present situation of women in Cuban society and clear analysis of state and party policy on the subject see Espin Guillois (1986).

19. Some recent examples of these interactions are the meetings between members of the Council of Ministers and Party leadership (including President Fidel Castro Ruz) with staff from Havana's hospitals, peasants from cooperatives, administrators and worker representatives from industries in Havana, conventions of students from different levels of the educational system, nationwide representatives of the Federation of Cuban Women, and the scheduled National Assembly sessions.

20. Kark (1981).

21. Mahler (1983).

REFERENCES

Aldereguia, J., "The health status of the Cuban population," *International Journal of Health Services*, Vol. 13, No. 3 (1983).

Behm, H., "Cuba: La Mortalidad Infantil Segun Variables Socioeconomicas y Geograficas — 1974" (Costa Rica: CELADE, 1981).

Capote-Mir, R. E., "La evolucion de los servicios de salud y la estructura socioeconomica en Cuba," *Revista Cubana de Administracion de Salud*, Vol. 5, Nos. 2 and 3 (1979).

Castro Ruz, F., Closing Speech at the Fourth Congress of the Federation of Cuban Women, 8 March 1985 (La Habana: *Granma*, 11 March 1985).

Cochrane, A. L., "National health service expectation," *Lancet*, Vol. 2 (1983) p. 154.

Comité Estatal de Estadísticas, "Evaluacion de los registros de defunciones: 1974" (La Habana: 1980).

Danielson, R., *Cuban Medicine* (New Brunswick: Transaction Books, 1979).

Dirección de Demografia, "Evaluación en 1974 de los Registros de Defunciones" (La Habana: Comite Estatal de Estadisticas, February 1980).

Dirección Nacional de Estadisticas, MINSAP, Datos del Registro de Profesionales.

Departamento de Orientacion Revolucionaria (DOR), "I Congreso del Partido Comunista de Cuba. Tesis y Resoluciones" (La Habana: Comite Central del Partido Comunista de Cuba, 1975).

Escalona Reguera, M., "El sistema nacional de salud en Cuba," *Temas de Administracion* (La Habana: IDS, 1983).

Espin Guillois, V., "La batalla por el ejercicio pleno de la igualdad de la mujer," *Cuba Socialista*, Vol. 6, No. 2 (20) (marzo–abril 1986).

Espinel Blanco, J. A., "Sistema de estadísticas de defunciones y nacimientos en Cuba," (La Habana: Instituto de Desarrollo de la Salud, 1981).

Gaceta Oficial de Cuba, Ley de reforma salarial (1980).

Gilpin, M., and H. Rodríguez Trias, "Looking at health in a healthy way," *Cuba Review*, Vol. 7, No. 1 (March 1978).

Granma (16 and 17 October 1985; 13 June 1986; 2 and 3 July 1986).

Grines, D. S., "The national health service expectations," *Lancet*, Vol. 3 (1983), p. 1073.

Grundy, P. H. et al., "The distribution and supply of Cuban medical personnel in Third World countries," *American Journal of Public Health*, Vol. 70 (1980), pp. 717–719.

Gutiérrez Muniz, J. A., et al., "The recent worldwide economic crisis and the welfare of children: The case of Cuba," *World Development*, Vol. 12, No. 3 (1984), pp. 247–260.

Hill, K., "An evaluation of Cuban demographic statistics, 1938–1980," in P. E. Hollerbach, and S. Diaz Briquets (1983).

Hollerbach, P. E. and S. Diaz Briquets, "Fertility determinants in Cuba," Committee on Population and Demography, Report No. 26 (Washington, DC: National Academy Press, 1983).

Instituto de Desarrollo de la Salud, *Sistema Nacional de Salud en Cuba* (La Habana: IDS, 1977).

JUCEPLAN, *Censo de Población y Viviendas 1970* (La Habana: Editorial Orbe, 1975).

Kark, S. L., *The Practice of Community-oriented Primary Care* (New York: Appleton-Century-Crofts, 1981).

López Valdés, J., "Organización de la asistencia pediátrica en el mutualismo," *Revista Cubana de Medicina* (January-February, 1966) p. 81.

Mahler, H., Speech presented at the conference on "Health for all in the year 2000" (La Habana: July 1983).

MINSAP, *Diez años de Revolución en salud pública* (La Habana: Editorial Ciencias Sociales, 1969).

MINSAP, *Organización del sistema único de salud* (La Habana: 1970).

MINSAP, *Fundamentación para un nuevo enfogue de la medicina en la comunidad* (La Habana: 1977).

MINSAP, *Annual Reports* (1980, 1982, 1983, 1985).

Mosley, W. H., "Will primary health care reduce mortality? A critique of some current strategies with special reference to Asia and Africa," paper presented at the IUSSP Seminar on Social Policy, Health Policy and Mortality Prospects (Paris: IUSSP, 1983).

Navarro, V., "Health, health services and health planning in Cuba," *International Journal of Health Services*, Vol. 2, No. 3 (1972).

Navarro, V., "Health Services in Cuba," *New England Journal of Medicine*, Vol. 287, No. 19 (1972).

"Primary Health Care," Report of the International Conference on Primary Health Care, Alma Ata, USSR, Sept. 1978 (Geneva: WHO/UNICEF, 1978).

Puffer, R. R., "Informe acerca de la calidad y cobertura de las estadísticas vitales en Cuba," Mimeo (Washington, DC: PAHO, Ref. AMRO No. 3513, 1974).

Rios Massabot, E., *Organización y descripción de los sistemas de estadísticas contínuas* (La Habana: Instituto de Desarrollo de la Salud, 1977).

Rodriguez Abrines, J., "La experiencia cubana del médico de la comunidad," Paper presented at the Third National Congress of Hygiene and Epidemiology (La Habana: 1985).

Salimano, G. et al., "Health for all in Cuba — Policies and strategies," Unpublished manuscript (New York: Center for Population and Family Health, Columbia University, 1984).

Stark, E., "Overcoming the diseases of poverty," *Cuba Review*, Vol. 8, No. 1 (1978), pp. 23–28.

Stein, Z., and Susser, M., "The Cuban health system: A trial of a comprehensive service in a poor country," *International Journal of Health Services*, Vol. 2, No. 4 (1972) p. 551.

Ubell, R. N., "High-tech medicine in the Caribbean," *New England Journal of Medicine*, Vol. 309, No. 23 (1983).

APPENDIX: NOTE ON THE QUALITY OF CUBAN HEALTH STATISTICS

Whenever evaluations or comparisons of the health status of a population are made, the first consideration must be the quality of the data. Cuban vital health statistics — those pertaining to births and deaths — are of excellent quality. Morbidity statistics on reportable diseases are also reliable because of the complete coverage of the population by health care services. The statistics which suffer from incomplete reporting are those for non-reportable diseases and those not included until now in strict follow-up protocols (silent morbidity like hypertension and risk factors like obesity or high serum cholesterol levels).

Cuban mortality data have been verified directly by international investigators (Puffer, 1974 and Behm, 1981) and by Cuban agencies (Comité Estatal de Estadísticas, 1980). Indirectly, Hill (1983) examined fertility, migration and mortality data comparing vital continuous statistics with census and survey information.

Puffer recommended to WHO that Cuban statistics be accepted as reliable, since the register was virtually complete and most of the international recommdendations for a reliable statistical system were in place in Cuba, including specialized training of personnel.

In 1974 Behm confirmed infant mortality rates based on registered births and deaths by using survey data.

Hill (1983, p. 232) found "Cuban demographic data to be of very high quality. Even the migration data, the Achilles heel of almost all demographic data systems fit well into the overall picture."

The only discrepancy was that between a 1979 survey and reported statistics. Hollerbach (1983, p. 177) accepts the official statistics "for two reasons: First, there is statistical consistency in the previous mortality estimates, and seond, there is no evidence that the official system has sharply deteriorated. Therefore, it appears that the survey data are probably distorted and that the official figures should be accepted as broadly accurate."

An indication of the "truthfulness" of Cuban statistics are the reported increases in rates for certain diseases and conditions, for example, infant mortality for certain years, suicide, syphillis and gonorrhea (see Table 1).

The author has been able to observe personally the care and zealousness with which mortality information is handled at all levels, and the strict observance of procedures relating to the definition and transmission of health-related data.

·8·

Worker Incentives in Cuba

ALEXIS CODINA JIMÉNEZ

1. INTRODUCTION

The harmonious combination of material and moral incentives constitutes one of the more complex problems in the construction of socialism. Analyzing the positive and negative experiences Cuba had undergone in this sphere, Fidel Castro indicated the following in 1973:

> Material incentives must be used together with moral incentives, without abusing one or the other, because the latter would lead us to idealism while the former would develop individual egoism. We must act in such a way that economic incentives do not become the exclusive motivation of man and so that moral incentives do not become the pretext for some to live off others.[1]

In Cuba, the mechanisms for moral and material incentives have evolved according to the different stages of the revolutionary process. This process has been influenced by objective factors such as the country's level of development, the conditions imposed by the economic blockade, and the constant threat of aggression from a powerful neighbor to which Cuba has been subject for the past 25 years. In addition to these objective factors, subjective factors, such as the inexperience of the first few years and the erroneous concepts applied in certain periods, have also played an important role.

The principal experiences and characteristics of Cuba's utilization of material and moral incentives may be analyzed through the stages that are presented below.

2. THE FIRST EXPERIENCES: 1961–66

The first efforts to organize labor and wages were undertaken by the Comandante Ernesto "Che" Guevara beginning in 1961. As Minister of Industry he promoted the presence of specialists from socialist countries — especially from the USSR — in Cuba. Under his direction, the training of thousands of former workers as technicians in these materials was begun; the basic elements of these methods were studied and divulged; and the first experiences of working out and implementing work norms and labor organization in industries that had come under the administration of the Cuban state were developed.

With the creation of these conditions, efforts toward the implementation of the first system of wages on a national scale were begun. The principal problems presented by the systematizing of the nation's wages may be summarized by the following: widely differing wages, disproportionate wages between branches, distinct names for the same job or imprecise job descriptions and qualification requisites, different systems of pay, and finally, the absence of any labor norms.

A detail that emphasizes the irregularities of the prevailing system was that in the sugar, nickel, and metallurgy industries the average wage was between 120 and 140 pesos per month — approximately half of the wage being paid in industries manufacturing cigarettes, beer, and soap. These discrepancies did not correspond to the complex technology utilized by these branches or to the necessities of the Cuban economy.

The following principles constituted the base of the wage system as its preparation was initiated in 1962:

---

*Translated by Lisa Galindo.

— Imposing one wage scale for the whole economy, according to the principle of equal pay for equal work, independent of the economic sector or branch.
— Stimulating the increase in the qualifications of the workers.
— Establishing work norms in all activities and centers of labor.
— Linking the payment of wage to the fulfillment of the norm. Any worker not meeting these requirements received only the proportional part of his/her corresponding salary. Premiums were paid for any production over the required norm (for every 1% of overproduction 0.5% of the wage was awarded, the maximum limit being fixed at the wage corresponding to the scale's highest category).
— Providing additional wage increments when work was carried out under harmful or dangerous conditions.
— Maintaining wages at old levels in cases where wages were greater than those defined by the new wage system. This was done because the disproportion was the result of gains won by the workers in their struggles against their employers — an accomplishment that the Revolution was compelled to respect. Workers in this situation received the wage that corresponded to them in addition to a bonus called "*plus*."

The integrating elements of the wage scales system initiated at that time were these: scales, tariffs, qualifiers and forms and systems of payment. The wage system classified the distinct types of labor performed into different groups according to the grade of complexity, and the level of qualification required. The scale relied on two elements: the number of groups and the coefficients. The number of groups determined the different grades of complexity while the coefficients expressed the quantitative relation between the said grades of complexity.

The scale that was established for workers at that time had eight groups. The coefficient ratio between the top and bottom groups was 3.08 to 1; that is, the types of work placed in group VIII were considered to be three times the level of complexity assigned to those in group I.

Wage tariffs defined the income that the worker should receive for a given unit of time (one hour) in the execution of a determined job. The tariffs differed from the scale coefficients, which only took into account the grades of complexity, in that they also took into account other factors such as the inherent noxiousness and danger in a job. Two types of tariffs were established to differentiate conditions: the basic tariff which was applied when the labor took place under normal conditions, and a tariff which took into account abnormal conditions. For example, if performed under normal conditions, a group II job had an hourly tariff of 0.56 pesos; if it was performed under certain abnormal conditions the tariff was 0.67 pesos. The basic tariffs established went from 0.48 pesos per hour for group I, which corresponded to the minimum wage, up to 1.49 for group VIII.

The *qualifiers* defined the characteristics of a given job, the knowledge of skill requisites needed to fill the job, and the group level of the job. Two types of qualifiers were established: those for jobs common to different branches of production such as chauffeurs, mechanics, etc., and those for activities performed only in a certain branch.

Finally, the forms and systems of payment established the methods that were to be applied to wages. Two forms of payment were applied: by time and by output. The first, which was the predominant form, was employed for activities whose characteristics made it impossible to determine norms of production or time, or whose fulfillment fundamentally depended on teams. Hourly, daily, or salary systems were established as forms of payment according to the characteristics of the work. The second form of payment, according to output, was employed in jobs where norms of production or time could be established, where the worker's duties were determinable, therefore making it possible to control strictly the quality and quantity of the work accomplished. The systems applied under this form of payment included, among others, piece rate and time rate with bonuses.

In order to realize the payment of wages in any form or under any system, it was necessary to possess a criterion for measuring the results that each alternative produced. *Labor norms* provided this function. Through different measurements and statistical data the volume of production or services that a worker in each job position should be producing under normal production and with a certain level of technology and production organization was established.[3]

The fact that 72.7% of all workers were located in groups I to IV — where jobs were the least complicated and required the least amount of skill — reveals two characteristics: first, the insufficient level of technological development that prevailed in most sectors of the Cuban economy, and second, the low cultural and technical level of the country's labor force. This latter fact is substantiated by the illiteracy rate at the time of the triumph of the Revolution which was over 25% for the adult population.

Taking into account this situation and the plans for industrialization that were already being designed, a chief objective of this new wage system was to increase workers' qualifications. The fundamental criterion for raising a worker's wage level was his graduation into a higher group on the scale which necessarily entailed raising his level of qualification. In accordance with this, the premiums for overproduction of labor norms a worker could receive were stopped when his total income reached the wage level of the group immediately above.

In order to initiate the experimental implementation of this wage system, 24 large centers of production were selected from various sectors of the economy which included industry, construction, transportation, agriculture, and fishing. Before applying the experimental system, its objectives, the plan by which it would be implemented in each specific center, and production quotas were discussed at workers' assemblies at the chosen centers.[4]

The results obtained from this experimental implementation were highly positive. The volume of production increased by 9.2%; the production per peso of wages paid increased from 9.10 pesos before the implementation of the system to 10.30 pesos; and labor productivity grew by 15% while the average salary grew by 3.5%. In addition, the 3.0% of workers in these centers who turned out to be in excess were relocated to other places.

After these initial experiences and after organizational papers had been prepared, the implementation of the wage system was gradually extended to the rest of the country beginning in 1963. By the end of 1966, it had been implemented throughout the productive sphere.

As a parallel development and as part of perfecting of the countrywide wage system, work examining the methods of labor organization was begun and the premises for what was later called the Scientific Organization of Labor were developed. The creation of this organization represents the applied techniques and methods of various disciplines and has as its component elements and objects of study the problems related to: the division and cooperation of labor, work methods, the organization and functions of jobs, work conditions, the norming of labor, the organization of salary, and work discipline.[5]

The implementation of the wage system at a national level meant raising the wages of 74% of agricultural workers, and in a more general sense, the wages of 25% of the country's labor force. Applying this unique wage system to the entire country signified an important step: it stimulated the increase of worker skills; it increased labor effectiveness and worker stability; and, it eliminated the chaotic situation presented by the existing countrywide organization of wages. The system ensured that workers in comparable occupations received the same basic wage. It eliminated sectoral and territorial disproportions that had responded neither to the complexities of labor nor to the needs of the country but, instead, to the particular interests and opportunities of the most profitable industries.

Notwithstanding the above, the stimulating effect of this system was limited. The first limiting factor was the maintenance of "historical wages." These were the wages of workers whose salaries were superior to those defined by the new system and which were respected by the Revolution. This was a just measure. From a political point of view, it would have been a mistake to lower the wages of these workers and thus lower their standard of living. This was not in keeping with either the interests or objectives of the Revolution.

Yet, from an economic point of view, it was not possible to establish unique wage scales and tariffs for these workers' wages. Such an action would have meant a general increase in wages which the economy could not have withstood. Furthermore, the establishment of scales and tariffs by branches, taking into consideration the wage level that each had at this time, would have entailed maintaining the same inequalities the planners wanted to eliminate.

The criteria applied by this new system resulted in 70% of non-agricultural workers receiving the "*plus*" or historical wage. In deciding the application of the wage system it had to be kept in mind that this situation would weaken the stimulating effect of the wage. However, given the incorporation of new workers and the openings created by those retiring, this was the only way of guaranteeing that the country's wage system would maintain a certain rationality and receptivity to the original objectives in later years. The situation that the Revolution encountered was the natural consequence of a longstanding absence of any uniform wage policy. There was no way the Revolution could rectify this situation immediately. By 1973 the proportion of workers receiving an historical wage had been reduced to 25%.

Another limiting element presented by the applied system was what was referred to as payment according to fulfillment of the norms. For every 1.0% of failure to meet the norm, the wage was reduced by the same proportion. For every 1.0% over the production norm, wage was raised 0.5%, although this increase was paid only until the worker reached the level of the group immediately above his/her own. The rationale for this practice, according to the proposals of that time, was that this would stimulate an increase in qualifications, limit the use of material incentives, and thus create a lever for the construction of socialism (material incentives were considered a negative element in the education of the workers). In practice, this policy did not stimulate an increase in production or in labor productivity.

Another mechanism of material incentives carried out at that time was the creation of an enterprise fund out of a portion of earnings or on the basis of an actual decrease in the enterprise's projected losses. This fund was not to exceed 5.0% of the annual wage fund and could be used for the delivery of individual and collective awards for production above the norms, for temporary financial aid to needy workers, for the financing of their vacations, etc.[6] The creation of this fund was proposed for those enterprises that operated under the system of "self-financing" applied during that period to agriculture, external commerce, and other sectors. Under this system enterprises had only one bank account and covered their expenses with their earnings.

Industries in other sectors (industry, construction, etc.) operated under a system called "budgetary financing." Under this system companies operated with an income account, the sum of which was returned to the state budget, and an expenditure account for wages, operations, and investments, which was decided by the limits of the budget approved for the enterprise. In actual practice, the concept of the enterprise fund did not succeed and the mechanisms of material stimulation were limited in these years to the wage system that had been implemented in 1963.

During this period, moral incentives were evident in the display of the intense and massive activity of an educational and practical nature that centered on voluntary work and socialist emulation. These elements constituted an extraordinary political value to the development of a socialist consciousness and of a new attitude toward work.

The role played by voluntary work in producing this new attitude in the masses and thus mobilizing them to achieve great tasks had its roots in the first "Red Saturdays" organized in Moscow by railway workers shortly after the Bolshevik Revolution. These initiatives, highly valued by Lenin and later extended to other regions and sectors of the country, proposed that such activities should plant the seeds for the work attitude that would characterize the society of the future: the attitude that one should work not for personal benefit, but for the common good.[7]

The promoter and developer of these ideas in Cuba during those years was Comandante Ché Guevara. From his high positions in Party and government leadership, he did not limit himself merely to preaching these concepts. He also set a personal example by actively and systematically participating in the mobilizations that took place at that time.[8]

The other element of moral incentives developed at that time was socialist emulation. At the First National Meeting of Emulation Fidel Castro defined the objectives as the following:

> Emulation must do the justice of exalting the values of our people, but it must fundamentally contribute to the optimization of the organization of labor; it must struggle for the adoption and general application of the vanguard experience, and of the most productive work forms or habits.[9]

In order to obtain the recognition and titles that were awarded in socialist emulation (achiever, winner, vanguard worker, work hero, etc.), goals were established for each worker or collective. The achievement of these goals required, by necessity, an outstanding effort. In keeping with the emulation system in effect at that time, goals for individual emulation were set according to the following guidelines: production plans or norms, savings, production quality, attendance and punctuality, professional excellence, and voluntary work. The following standards were added for the case of collective emulation: reduction of accidents in the workplace and the total number of workers that were involved in collective emulation. Results obtained in striving toward these goals were graded by points and the winners determined by the total of points.

Socialist emulation played an important political and economic role during this period (1962–66). Above all, it was instrumental in mobilizing the masses for the achievement of specific economic objectives such as the fulfillment and overfulfillment of production plans, the saving of materials, and an increase in production quality. Emulation also positively influenced work discipline since a worker who was frequently tardy or absent was obliged to account for these lapses to his/her labor collective, which in turn affected the worker's possibilities of receiving the recognition of vanguard worker by the collective.

Nevertheless, the emulation system of this period suffered from certain defects that limited its ability to mobilize the masses. Among these were an excessive rigidity that limited the development of the collective's initiative, the excessive number of indicators, and a certain formalism in the selection of the most outstanding workers.

For these reasons, new norms were adopted at the end of this period. These were designed to make the procedures of socialist emulation more flexible, to encourage the development of worker collectives' initiative, and to elevate the effectiveness and mobilizing capacity of this important instrument of moral incentives.

3. THE SUPPRESSION OF THE MECHANISMS OF MATERIAL INCENTIVES, ERRORS, AND CONSEQUENCES (1967-70)

Beginning in 1966, proposals were made in the Cuban press and by leaders of the Revolution concerning the contradictions that had resulted from the monetary and mercantile relationship among state enterprises and the basic concepts behind a socialist economy, as revealed by the negative influence played by material incentives in the creation of a socialist and communist consciousness.

Carried away by the enthusiasm produced by labor's successes and by the exemplary attitudes of several worker collectives, in those years we thought that all workers had already acquired a consciousness and attitude toward work that made it unnecessary and detrimental to use material incentives. For instance, a resolution passed by the Labor Ministry indicated that

> the advances made by our Socialist Revolution have clearly demonstrated that man is capable of realizing truly productive feats without requiring the application of wage forms in which the increase in productivity carries with it a higher wage. This reveals that a whole series of systems and forms of payment by output have become ultimately detrimental to the development of a communist consciousness.[10]

In keeping with this conception, after 1967 monetary-mercantile relations in state industries were eliminated, studies tied to the norming and organization of labor were abandoned, wages were frozen, and the use of material incentives was discontinued entirely.

Once the idea developed that all workers possessed a heightened work consciousness, controlled work schedules were substituted by what were called "consciousness schedules." In addition, a resolution was promulgated giving collectives the authority to decide that the best workers would receive 100% of their wage when they missed work for justifiable reasons and when they retired. This measure, which was legislated for special cases, was generalized in actual practice and applied extensively. In addition, payment of overtime work was eliminated.

In an effort to reduce paper money a "gratuity" policy was initiated which consisted of eliminating the payment for certain services and of instituting the gratuitous delivery of material goods. At the beginning of this stage — in 1967 — it was contended that people would no longer have to pay for rented housing by 1970. Furthermore, an "egalitarian" principle started to be applied to the payment of wages: in other words, labor was remunerated without taking into account job complexity or the results obtained.

These policies coincided with an increased share of national income going to investment. At the same time, there was widespread popular mobilization to guarantee the production of the 10 million tons of sugar projected for 1970 (the average for the 1961-65 quinquennium had been 5.2 million).[11]

At the end of the 1970 sugar harvest, huge disproportions and other negative phenomena were evident in the Cuban economy. Many investments were paralyzed, the production of non-sugar agricultural industries remained at their 1965 level, and the production of the majority of industrial branches, with the exception of the sugar industry, had suffered significant deterioration.

The decrease in the availability of merchandise on the national market reduced per capita consumption (social and personal) to 91% of its 1961 level. This, combined with the effects of the gratuity policy and the discontinued practice of linking wage to output, produced a significant increase in the amount of money in circulation. Accumulated liquidity (money in the hands of the population) grew from 1.525 million pesos in 1966 to 3.335 million in 1970, a figure somewhat higher than the total of wages paid out in one year.[12]

This situation presented a strong disincentive to labor that manifested itself in the form of a decrease in labor productivity and an increase in absenteeism. In 1966, for each peso in wages paid, a production of 1.58 pesos was obtained; in 1970 this decreased to 1.38. In many labor centers worker absenteeism reached 20% and the use of the workday did not surpass 60% of efficient levels.

In the Central Report to the First Congress of the Cuban Communist Party (PCC) held at the end of 1975, Fidel Castro made a profound critical analysis of the errors committed during this stage indicating:

> In the management of our economy we have suffered from errors of idealism. . . . By idealistically interpreting Marxism and leaning on the practices consecrated by other socialist countries, we sought to establish our own methods. . . . When

it seemed that we were approaching communist forms of production and distribution, in reality, we were straying from the correct methods of laying the foundations for socialism.[13]

4. THE BEGINNING OF THE PROCESS OF RECTIFICATION AND THE RE-ESTABLISHMENT OF MATERIAL AND MORAL INCENTIVES (1971–75)

At the end of the 1970 sugar harvest, Castro analyzed the problems the national economy had faced and made a plea to the people to work to raise the efficient use of labor and material resources. Measures were adopted to stimulate an economic recovery and to begin to rectify the consequences of errors committed.

In the area of labor and wage organization a set of efforts was initiated to achieve the following immediate objectives:

— achieving a major utilization of human resources, incrementing productivity, and minimizing cost;
— establishing in all work centers elementary labor norms (ones based on experimental statistics) and adjusting personnel staffs;
— creating a collective awareness of the economic, social, and political importance of labor and wage organization;
— preparing technical experts in the area of work organization; and
— strengthening wage and labor departments at all levels.[14]

In the years 1970–73, with the cooperation of government agencies, unions, the Ministry of Labor, and the Party, seminars on the problems suffered by the economy were offered to over 100,000 directors and technical workers; thousands of worker assemblies were held; and various materials were published with the intention of increasing workers' knowledge of these issues.

Educational institutions were created from which by 1973 more than 3,000 time and motion experts and close to 500 technicians graduated. By the end of 1973, these accomplishments allowed for the establishment of elementary labor norms for close to two million workers — approximately 82% of the work force in the state sector — and for the adjustment of the staffs of almost 2,000 production or service units.

The results obtained from these measures were highly positive. In a study conducted in 500 work centers, productivity grew an average 21% in three years and employment was reduced by 6%.

The 1973 celebration of the VIII Congress of the Cuban Workers marks an important moment in the development of labor and wage organization and in the re-establishment of material incentives. The theses, discussions, and principal resolutions that emerged from this Congress all centered on these problems.

Among the important agreements reached were: the reinstatement of the practice of paying wages according to the quantity and quality of work contributed by each worker; cancellation of the resolution that had made it possible to pay outstanding workers 100% of their wage when they were absent or retired; the gradual elimination of "historical wages"; revision of the wage scales, tariff qualifiers, premiums for abnormal work conditions; and the perfection of labor standards, which due to their elemental character often did not stimulate labor productivity. The theses and resolutions were discussed and analyzed by all workers prior to the Congress. The importance of this Congress, hence, lies not only in the resolutions adopted but also in the fact that preparations for the Congress constituted an important step in the education of workers. The workers were thus informed of the mistakes that had been committed and of the policies that would be followed in the national sphere of labor and wage organization in subsequent years.

Among other decisions made in 1973, the National Center for the Scientific Organization of Labor was created and ascribed to the Ministry of Labor (presently called the State Committee of Labor and Social Security). The Center was entrusted with the task of developing the methodological base for the scientific organization of labor. As a result of the work done during 1974 and 1975, new qualifiers were introduced into many branches of the economy; wage tariffs for workplaces with abnormal conditions were increased; and systems of payment according to output were implemented for close to a million workers.

Although these constituted important measures, they were still insufficient. In order for them to wield a positive influence on the increase of labor productivity, it was necessary to arrest the tendency of money in circulation to increase — as had progressively been the case during the previous quinquennium — and to retire the largest amount possible of this money. Among the several measures adopted were: increasing the production of agriculture and other branches of industry that produced consumer goods (textiles, shoes, food); and increasing the importation of durable consumer goods such as televisions, refrigerators, and beginning the national production of such goods. Also, prices of luxury items such as cigarettes, beer, and alcoholic beverages were increased.

As a result of these measures, starting with the second half of 1971, the amount of money in circulation began to be reduced. By the end of 1971 the amount was reduced by 150 million pesos; in 1972 680 million pesos were withdrawn from circulation; and in 1973 more than 400 million pesos were taken out of circulation. In two and a half years, a total of more than 1,300 million pesos had been withdrawn, achieving an acceptable financial balance.

The relatively positive behavior of the Cuban economy during the 1971–75 quinquennium further reflected the results of measures taken. The Global Social Product grew at an average rate of a little over 10%. Labor productivity grew an annual average rate of 10% while the average wage grew by 3.6%.[15]

At the end of 1975, the entire revolutionary process of economic development in Cuba was assessed at the First Congress of the Cuban Communist Party. A thorough analysis was made of the errors committed in the management of the economy and guidelines for economic and social development were traced for the 1976–80 quinquennium.

5. THE FIRST CONGRESS OF THE PCC AND THE 1976–80 QUINQUENNIUM

One of the resolutions that was approved at the First Congress of the PCC called for the establishment of a new system of Planning and Management of the Economy. This system proposed the utilization of financial categories and mechanisms such as the budget, bank credit, earnings, etc. Some of the more important objectives of this system were:

— Attaining the economy's maximum efficiency through the most rational use of productive resources (material and human) and producing a maximum of results with a minimum of spending.
— Establishing mechanisms that would assure the necessary work discipline, which would contribute to a consistent increase in productivity and stimulate a rise in the quality of production of goods and services.
— Establishing an adequate balance between material and moral incentives.[16]

It was argued that material incentives offered directly to the workers should be fundamentally carried out by the consistent application of the socialist principle of distribution. In addition, it was proposed that the system of incentives should consider granting collective incentives that would be contingent on the economic efficiency of the enterprises. To this end, stimulation funds were created with a portion of the enterprise's profits, according to the enterprise's fulfillment of various indicators. These funds would be used both for individual material rewards to the workers and for the improvement of their sociocultural conditions.

The Directives for the Economic and Social Development of the 1976–80 Quinquennium — also approved at this Congress — proposed "continuing the process of tying norm fulfillment with the wage and advancing to the application of the principles of the scientific organization of labor, the technically based norms, and other rationalizing measures."[17]

In the area of wage and labor organization, the following was accomplished: systems which paid by output were applied to close to a million workers; new qualifiers were applied in various branches and remuneration for abnormal job conditions was generalized; and bonus systems were introduced into different industries to more than 500,000 workers. Furthermore, more than 1,500 technicians were trained in the organization of wage and labor and legislation prohibiting the creation of new historical wages was promulgated.

Nevertheless, there were many problems which indicated that things were not going well in this sphere. The plan to link wage to work was not being effected in many enterprises; work norms were being surpassed by large degrees which revealed their low quality and their unmotivated character. In addition, the application of the principles of the scientific organization of labor was not advancing. In general terms, the behavior of the indicators of labor productivity — average wage, the cost of wage per peso of production — were not as favorable as planned.[18]

A study of these work problems revealed that one of the primary contributing factors was that the small technical force which existed in the country had been concentrated in large projects involving the "scientific organization of labor." In addition to the fact that these projects had not contributed to the average level of development in the country, they had limited the possibilities for working on less complicated projects that would have been more viable and would have embraced a greater number of productive centers.[19]

After reviewing the experiences of the previous years it was decided to follow a policy of what was called the "Basic Organization of Labor," which was more technically advanced than the "Elementary Organization" that had been reinitiated in 1972–73. While more advanced, it was not as complex as the scientific organization of labor which had required for its extensive application, highly qualified technicians and conditions of a general organization and a technical economic base that did not exist. While the scientific organization of labor was not entirely dismantled, it was limited to those industries that possessed the conditions to make its application feasible.

Independent of the objective and subjective factors that affected wage and labor organization during this period, there was the more general problem of the current wage system. This system had been conceived in 1962 and the conditions under which it had been implemented had substantially altered its character. In addition to changes produced in the Cuban economy by the considerable volume of investments made in the years following the triumph of the Revolution, the educational levels of the Cuban population had undergone significant transformations. According to estimates, in 1958, 86% of all workers had not continued their education past the sixth grade and of those, 29% were illiterate. Of the remainder only 8.1% had completed secondary school and 3.7% had either a semi-specialized or higher education. By 1979, 51% of all workers had attained a primary school education, 25% a secondary school education, 8.3% a pre-university level of education, and 9.4% a semi-specialized or higher education.[20]

In fulfillment of the First PCC Congress resolutions concerning the implementation of the SDPE, the experimental creation and distribution of stimulation funds was begun in 191 enterprises in different economic sectors. In 1980, this number was increased to 204 enterprises.

The stimulation funds were created in both profitable and non-profitable industries. Because the prices at which non-profitable enterprises sell their products — products of first necessity such as meat and milk — are lower than their costs of production, these industries are financed by the State. In profitable industries, funds are created with a portion of the industries' earnings and in conjunction with their performance in select economic indicators. The indicators used at that time included the increase in labor productivity and the reduction in cost per peso of output. In non-profitable enterprises, stimulation funds are created from a decrease in the enterprises' state subsidy. In both types of enterprises, the resources accumulated in these funds are designated for the award of money prizes to workers or for sociocultural measures such as housing repairs, worker cafeterias, construction of beach cabins, etc.

The first results obtained by the funds were positive. Industries participating in the experiment increased their production indicators and labor productivity. In addition, the level of decrease in their costs of production was better than that achieved by the rest of the country's industries.

6. THE GENERAL REFORM OF WAGES, THE PRICE REFORM, AND THE ORGANIZATION OF WORK BRIGADES, 1980–86

The General Reform of wages, implemented in mid-1980, had the following basic objectives:

— Ensuring that the wages paid corresponded to both the quantity and quality of the work done by each individual.

— Stimulating an increase in labor productivity, raising the workers' level of qualifications, decreasing costs of production, and creating conditions which favored improvement in labor organization and discipline.

— Benefiting lower-income workers by raising the minimum wage; increasing also the highest wages so as to promote the development of technical workers of the highest caliber as well as to motivate the enterprise directors and skilled personnel. Under the General Reform, the highest wage was 5.29 times the minimum wage; under the 1963 Reform, this same figure was 4.33.

— Establishing for technical personnel the principle of paying by the job itself rather than by the job title (engineer, economist, etc., as was done under the 1963 Reform).

— Using corresponding wage incentives to distribute better the labor force among the prioritized regions and branches of the economy.

— Differentiating the wages of management personnel according to the different functions and degrees of responsibility they assumed. The idea is established that a director has to possess a high degree of qualification, and that accordingly, as a general rule, the director should receive a higher wage than other workers in an enterprise.

In order to implement the General Reform of Wages it was necessary to: enlarge the groups on the complexity scale, raise wage tariffs, create more detailed standards and requisites for technical and managerial jobs; perfect work norms; establish and extend the system of payment by output to all normable activities, and develop further bonus and stimulation fund systems.

The Reform produced significantly positive results from its very first moments of implementation. By the end of 1981, more than

90% of all workers received the wage accorded to them under the new reform. Labor productivity grew about 10% while the average wage grew by 4.5%. That year, 85% of the increase in production was obtained through an increase in labor productivity (the other 15% was produced by an increased number of workers).

The wage reform caused a significant increase in wage expenditures. Through December 1981 its implementation represented 535 million pesos more in wages. Subtracting the savings made — produced by the application of measures reorganizing labor and reducing jobs (calculated at 194 million), among others — from this figure, the net cost of the wage reform may be estimated at 341 million pesos.[21]

The minimum wage increased by 14%. The average monthly wage increased significantly in some sectors: 30% in the agriculture; 26% in the sugar industry; 14.5% in construction; and 13.7% in the industrial sector as a whole.

This situation had been anticipated. From the time the Reform was conceived it was known that the increase it would generate in the wage fund could not be financed — at least in the first years — with the production increases produced by the increased labor productivity. When the Reform was initiated, the population was informed that a rise in the price of certain products that had been frozen since 1962 might be necessary; prices had been frozen even though their cost of production or importation had increased substantially during the intermittent years.

A retail price reform was carried out at the end of 1981. There was a subsequent price increase affecting close to 10% of the products available to the population at that time. Among the products affected by the reform were: pasteurized milk, the price of which rose from 0.20 to 0.25 pesos per liter; rice, from 0.20 to 0.24 pesos per pound; beef, from 0.55 to 0.70 per pound; pork fat, from 0.25 to 0.30 per pound; potatoes from 0.07 to 0.12 per pound; and special gasoline, from 0.16 to 0.27 per liter. Textile prices also increased, as did hotel and restaurant prices.[22]

According to calculations made through family budget surveys by the State Statistical Committee for all provinces of the country, these price increases meant an increase in spending of eight pesos per month for the average four person family. Meanwhile, the average additional income caused by the wage reform for the same family was 36 pesos in 1981.

New forms of labor organization constitute an important aspect of the increase in labor productivity and the functioning of the mechanisms of material and moral incentives. The organization of labor into brigades was first begun during the late 1970s, but brigades were rapidly extended after 1980.

The experiences of other socialist countries have shown that the creation of work brigades produces a considerable economic effect: an increase in labor productivity and a decrease in losses and expenses caused by unproductive time. In addition, brigades have a great social significance in that they permit an enrichment of daily work life. By making possible the development of the collectives' initiative while concurrently creating new activities, the work brigades lead to greater worker participation in the direction of production.

Although the same general principles have been applied, the work brigades in the Cuban agricultural sector go by the name of Permanent Brigades of Production while those in the industrial sector are called Integral Brigades of Production. The integral or permanent brigade of production is considered the basic unit of production and the primary link of the labor collective to the entrprise.

A brigade is a stable group of workers. The workers are given the necessary resources for the completion of a certain task. They function according to the principles of internal self-financing (*cálculo económico*): that the value of what they produce or of the service they offer should always be greater than its expenses; that material and moral incentives should be utilized; and that the workers have the authority to organize their own labor.

The principal characteristics of a Cuban work brigade are the following:

— The workers are brought together to carry out a group job, the result of which constitutes a final product or a certain part of it (article, work, project, volume of loading and unloading, agricultural activity, etc.).

— A brigade works on the basis of an annual plan that all members are involved in drawing up. This plan must concur with the principal economic and production indicators and should be broken down into trimesters, and, at times, months or days, according to the characteristics of the brigade's activity.

— Each brigade is assigned the technical, material, wage, and human resources that are indispensable to the fulfillment of its productive tasks.

— The brigade's collective task is distributed among its members according to the division and cooperation of labor that they have established.

— A significant portion of the brigade's wage is tied to the final results of their work by which is understood the fulfillment of the planned indicators that define their activity. Wages are distributed among members according to their participation in the collective task.

— The brigade possesses a certain degree of operative autonomy in the execution of its task; the direction and overseeing of its achievement is provided by the brigade chief, elected by the brigade members.[23]

The purpose of organizing labor into integral production brigades should be to increase the volume and quality of production through the better organization of labor. Production brigades should also reduce the costs of production by effectively using the natural and human resources put at their disposal. They should increase the wage earnings of their members and thus make possible an improvement in the material and social well-being of workers. Finally, production brigades should raise the skills of workers and develop the characteristics of a communist attitude toward work.

Brigades are classified by complex and specialized divisions according to the qualifications of their members. The first divisions are those that cluster workers of different professions and organize them to fulfill jobs that are different, by their nature or by the technology of the productive process; in other words, heterogeneous albeit interrelated tasks are undertaken. Specialized brigades are essentially those that organize workers from the same profession to work in homogeneous technological processes whose volume and sequence permit the employment of many workers from one profession.

At the moment of actually forming the brigade, the workers' personal interests and characteristics are taken into the highest possible consideration so as to assure the harmonious integration of the collective. In addition, all members of the brigade are consulted about the incorporation of each member.

The principal instrument for the direction and control of the functioning of the work brigades is the plan that is established in accordance with the other objectives and tasks that must be achieved by the department, plant, enterprise, etc. Payment for the brigades' work is tied to the greatest extent possible to the final results of production. The wages brigade members receive are contingent on the complexity of the job, on the conditions in which it is done, and on the established work norms. The system of payment that is preferably applied is that of collective output, where the wage is set according to a predetermined unit of the finished product or service of a certain quality.

The new system of labor organization and stimulation creates favorable conditions for the further development of socialist emulation. The primary form this takes under the new conditions is that of collective emulation between the brigades of enterprise or industry.

Emulation among brigades is based on the quantitative and qualitative results of the task. Such standards include the increase in labor productivity, improvement in the quality of production, efficient use of resources, work discipline, decrease in accidents, development of initiatives, etc. Individual emulation is organized within the brigade and is directed by stimulating an increase in an individual's contribution to the collective's final product and to the development of the initiative of each of its members.

As in the case of other workers, brigade members are also eligible for prizes which are awarded from the stimulation fund created by the enterprise for this purpose. The size of each brigade's fund is calculated from the prize fund created at the enterprise level, taking into consideration each brigade's productive results and its relative weight within the company.

Up to 10% of the fund the brigade receives may be designated for individual prizes which are awarded to members with outstanding labor achievements directly connected to socialist emulation. Up to 5% goes to members who have completed productive tasks of special importance to the perfecting of production, introduction of new techniques or improvement in the quality of production or to workers who have maintained a flawless level of moral conduct. The remainder of the fund is distributed among all the brigade's members, taking into account each worker's basic wage and performance in various areas: fulfilling or overfulfilling work norms, increasing labor productivity, saving raw material, etc.[24] By 1985, more than 100 Permanent Brigades of Production had been created in agricultural enterprises while close to 200 Integral Production Brigades had been formed in the industrial sector.

The economic results obtained by the enterprises experimenting with the work brigades were superior to the average results obtained by the rest of the country's enterprises. A group of the former had increased labor productivity by 15% and production by 8.7% in one year.[25]

During this quinquennium new mechanisms that stimulated specific objectives had also been created. These objectives included the creation of funds for material incentives, as well as the establishment in enterprises of bonuses which were awarded to workers who accomplished increases in production, conserved fuel and electrical energy, recovered scrap iron or industrial waste that could be used as raw materials, or achieved other similar objectives.

The 1981–85 quinquennium turned out to be one of the most productive periods in the history of the Cuban economy. Real Global Social Product grew at an average annual rate of 7.3%, 5% higher than the figure projected by the Plan, while industrial production grew at an average annual rate of 8.8%. Labor productivity increased at an average annual rate of 5.2%.

The Second PCC Congress was held in February of 1986. The PCC's proposed program, which will be discussed by all workers in the course of this year, was among the important documents approved. The fundamental strategy orienting the country's economic and social development over the next several years will be outlined in these discussions.

Among the Economic and Social Guidelines approved by the Congress for the 1986–90 quinquennium were some related to the moral and material incentives for the next five years. Regarding these, it was proposed among other things that

> during the quinquennium economic incentives should become a fulcrum for enterprises to encourage workers to strive for greater efficiency . . . the application of the principle of linking production expenses to final results should constitute one of the primary avenues for strengthening, in the eyes of society, the authority of the basic economic decisions made on respective levels by each collective.[26]

Regarding the perfection of the SDPE in the sphere of labor, wages, and incentives, it was proposed that a serious effort should be undertaken to assure that wages be increasingly responsive to the increase in productivity. At the same time, wages should play an important role in extending the number of operations where payment by output was possible, as well as encourage the most intimate link between wage and sales.[27]

During the next quinquennium an important role in the perfecting of labor organization will be assigned to the extension of permanent or integral production brigades. For this, indispensable conditions must be created and ensured through adequate wage schemes and the incorporation of qualified personnel — into jobs directly related to production.

The gradual incorporation of mechanisms that stimulate a more efficient use of labor (reducing excess personnel) has also been proposed for the next quinquennium. These efforts should be initiated in deficient areas of the labor force. The norming of work and its extension will receive ever more attention over the next five years. Enterprises will be obliged to adjust and update the norms which should be implemented on the enterprise level in accordance with the characteristics of every job.

NOTES

1. Castro (1973), p. 29.

2. González (1975).

3. *Ibid*.

4. Martínez (1963).

5. González (undated).

6. Infante (1964).

7. Lenin (   ), p. 365.

8. Comandante Ernesto "Ché" Guevara's principal ideas on these issues are developed in various articles and speeches collected in the *Colección Nuestra América*, Tomo II (Guevera, 1970). The following, among other work, may be consulted: "Qué debe ser un joven comunista," pp. 161–175; "En la entrega de certificados de trabajo comunista," pp. 238–259; "La juventud y la Revolución," pp. 308–318; "Una actitud nueva frente al trabajo," pp. 332–350; and "El hombre y el socialismo en Cuba," pp. 367–384.

9. Cited in Rodríguez (1966), p. 97.

10. González, (1984), p. 167.

11. JUCEPLAN (1970, p. 35; 1974; p. 35).

12. JUCEPLAN (1971), p. 51.

13. Castro (1975), p. 107.

14. González (1984), pp. 171–189.

15. Comité Estatal de Estadística (1975 edition, p. 39; 1976 edition, pp. 39, 50).

16. Resolution regarding the system of the Dirección y Planificación de la Economía, appearing in Partido Communista de Cuba (1975), pp. 189–207.

17. Partido Comunista de Cuba (1975), p. 111.

18. JUCEPLAN (1980), pp. 287–311.

19. Speech given by Leonel Soto, member of the

PCC's Central Committee, at the II Encuentro Nacional de Organización del Trabajo (12 diciembre 1979).

20. González (1984), p. 189.

21. J. Benavides, speech given at the XV Congreso de la CTC, 22 February 1984.

22. *Granma* (14 December 1981).

23. Resolution no. 4390, State Committee of Labor and Social Security, "Principios generales para la organización y estimulación del trabajo en las Brigades Integrales o Permanentes de Producción" (1 June 1985).

24. Resolution no. 1124/82 of the Junta Central de Planificación, Reglamento General para la distribución del Fondo de Premios, Gaceta Oficial No. 19 (1 March 1982).

25. Roberto Veiga, CTC Secretary General, speech given at the National Seminar on New Brigades, held in 1985.

26. JUCEPLAN (1986), p. 8.

27. Proposed bill of resolution concerning the perfecting of the SDPE, p. 14.

REFERENCES

Castro, Fidel, "Discurso en el acto central en comemoracion del XX Aniversario del Asalto al Cuartel Moncada," *Economia y Desarrollo*, No. 19 (Sept/Oct 1973).
Castro, F., *Informe Central: Primer Congreso del Partido Comunista de Cuba* (La Habana: DOR del CC del PCC, 1975).
Comité Estatal de Estadística, *Anuario estadístico de Cuba* (La Habana: Junta Central de Planificación, 1975, 1976 editions).
González, L., *El Salario* (La Habana: Centro de Estudios Demográficos, 1984).
González, L., *El trabajo y su remuneración* (La Habana: 1975).
González, L., *Problemas metodológicos de la organización científica del trabajo* (undated).
Guevara, Ernesto (Ché), *Colección Nuestra América*, Tomo II (La Habana: Casa de las Américas, 1970).
Infante, J., "Características del funcionamiento de la empresa autofinanciada," *Cuba Socialista*, No. 34, (junio 1964).
JUCEPLAN, Dirección de Estadísticas, *Boletín estadístico de Cuba* (La Habana: JUCEPLAN 1970).
JUCEPLAN, Dirección de Estadísticas, *Boletín estadístico de Cuba* (La Habana: JUCEPLAN, 1971).
JUCEPLAN, Dirección de Estadísticas, *Boletín estadístico de Cuba* (La Habana: JUCEPLAN, 1974).
JUCEPLAN, *Segunda Plenaria Nacional de Chequeo de la Implantación del Sistema de Dirección y Planificación de la Economía* (La Habana: JUCEPLAN, julio 1980).
JUCEPLAN, *Lineamientos económicos y sociales para el quinquenio 1986–1990*, First Part (La Habana: Editora Política, 1986).
Lenin, V. I., *Obras escogidas en tres tomos*, ediciones en lenguas extranjeras, Vol. 3 (Moscow).
Martínez, S. A., "La implantación del nuevo sistema salarial en las industrias de Cuba," *Cuba Socialista*, No. 25 (octubre 1963).
Partido Comunista de Cuba, *Tesis y Resoluciones Primer Congreso del PCC* (La Habana: DOR del CC del PCC, 1975).
Rodríguez, B., "Las nuevas normas de la emulación socialista," *Cuba socialista*, No. 56 (abril 1966).

·9·

Power at the Workplace: The Resolution of Worker-Management Conflict in Cuba

LINDA FULLER

1. INTRODUCTION

Central to the work experience is a complex set of social relationships and interactions in which people engage during the day-to-day act of producing goods and services. Perhaps the most universal characteristic of these workplace social relationships in the contemporary world is a highly unequal division of power. Characteristically, a relatively few owners and high-level managers monopolize power at work, while the vast majority of workers possess very little.

As a socialist country, Cuba claims to have altered the distribution of power in society in favor of the working class majority. One place we should hope to see evidence of the truth of this assertion would be in efforts to modify the unequal division of power at the workplace in favor of the producers. An investigation of the distribution of power in the workplace social relations could adopt a variety of approaches. The kinds of resources available (or not available) to workers versus owners and managers — organizational, fiscal, educational, etc. — might be compared. The overall structure of the organization and how much power, responsibility and privilege attend to each level within it might be studied. The content of workers' jobs could be examined to determine the extent to which workers are in control or are being controlled as they expend their labor hourly and daily.

All such perspectives are important in forming a complete picture of how power is distributed at work, but none is precisely the approach taken below. This discussion of power distribution at the Cuban workplace will focus on the times when overt disputes erupt between workers and managers over issues of discipline and workers' rights. On these relatively rare occasions when conflict dominates over consensus and cooperation, the nature and the division of power at the workplace can often be seen most transparently. To understand what occurs when conflicts surface we will look at the mechanisms which have been devised to resolve workplace disputes: how are they structured and how do they operate? To what extent are they biased towards management-preferred outcomes, or, conversely, how much power do Cuban workers have to discipline themselves and to interpret and enforce the formal rules which define the basic conditions under which they work? Do they reinforce existing power inequalities, or do they foster the growth of control by the many?

Our investigation of conflict resolution in Cuba will span the entire postrevolutionary period. This is because workplace social relations — and the dispute resolution mechanisms that help define them — have undergone important changes over the last quarter century. Comparing how worker–management disputes have been handled at different times will help clarify the elements of the structures themselves which have particularly important effects on power distribution at work. It will also highlight some aspects of the environment in which the grievance systems operate which influence the power dimension of workplace social relations. In concluding we will direct our attention to one of these extrinsic factors — the unions — and look at how they affect workplace power through their connections with the conflict resolution bodies.

2. A TRIPARTITE SOLUTION: THE EARLY GRIEVANCE COMMISSIONS

The first six years following the victory over Batista were unsettled ones in terms of workplace dispute resolution. The period between 1959 and 1962 saw four separate pieces of legislation concerned with methods for resolving conflicts between workers, managers and employers. Such frequent changes meant that none of these systems ever functioned very satisfactorily, if they functioned at all at many worksites. The last in this series of procedures, established by the 1962 Law of the Administration of Labor Justice, remained in place for two years. As such it can be considered the first relatively permanent method of handling workplace conflict devised under the revolution.

The 1962 law set up Grievance Commissions (*comisiones de reclamación*) at the worksite level. Direct producers were a minority on these tripartite bodies composed of one worker delegate, one management delegate, and a chair representing the Ministry of Labor. Reports on how these three members were chosen are inconsistent.[1] The commissions had primary jurisdiction over conflicts involving some alleged violations of workers' rights. However, their power here was limited because they could not hear cases involving certain wage disputes or "the qualification, registry, distribution, selection and promotion of manpower."[2] The commissions also heard disciplinary cases: dismissals and appeals brought by absentees who had incurred punishments ranging from public warnings to transfers. Cases were decided by majority vote, and appeals were handled by five-person bodies appointed by and representing the Cuban Labor Confederation (CTC), the administration and the Ministry of Labor, although according to two sources, the Ministry of Labor retained final authority to issue decisions, which could not be appealed (CERP, 1963, p. 135; Hernández and Mesa-Lago, 1971, p. 220).

Assessments of how well the commissions functioned varied. Zeitlin (1970, p. 191) reported that they "had succeeded to a great extent in gaining the workers' approval," yet they were also criticized for supporting workers too often, as well as for the opposite fault of siding with management (Hernández and Mesa-Lago, 1971, pp. 220, 247; CERP, 1963, pp. 47, 134–6). An interesting aspect of the 1962 law which made access to the worksite grievance mechanisms easier and less costly for workers, many of whom were illiterate at the time, was the provision allowing cases to be presented orally.

3. WORKER CONTROL OVER WORKSITE CONFLICT RESOLUTION: THE *CONSEJOS DEL TRABAJO*

The Grievance Commissions were disbanded in 1965 when a new law was passed establishing the Work Councils (*Consejos del Trabajo*), which have played an important, though variable, role in worker–manager dispute resolution ever since. The most noteworthy change introduced by this law was the discontinuation of both administration and Ministry of Labor representation on worksite grievance bodies. Thereafter, throughout the 1960s and 1970s, only workers would judge alleged cases of indiscipline and violations of labor legislation at Cuban workplaces. The 1965 law called for five *Consejo* members to be selected in a secret ballot election at each worksite. Council members were elected for three-year periods, although many apparently seek, and are regularly elected to consecutive terms: the job of a Council member can be a delicate one, not to everyone's liking. One teacher explained to me that in his work center a person could be a member of the *Consejo* as long as she or he cared to, because "Nobody really *wants* the job. But," he added quickly, "everyone recognizes that it's better than having the boss do it."[3] Starting at least by the early 1970s, *Consejo* members received training on labor law and the role of the Councils at a National School of Labor Judges run by the Ministry of Labor. Such training was felt to be necessary for workers on the Councils to be able to make judgments independently, without the interference of management or political cadres.

The *Consejos* were empowered to resolve conflicts between workers and administrators over issues of discipline and workers' rights. Absenteeism was the biggest discipline problem in the 1960s and 1970s, and available evidence indicates that much of the Councils' time was spent on cases of this sort. Other violations of discipline heard by the worksite grievance bodies included tardiness; failure to meet work quotas or time schedules; disobedience; negligence; lack of respect for superiors, fellow workers and visitors, or mistreatment of their work; physical offenses; damage to work center equipment or property; fraud; and robbery. In 1972–73, 83% of the cases before the *Consejos* involved indiscipline, and though this percentage decreased as the decade progressed, it is clear from the union newspaper that cases involving such things as absenteeism, worker violation of health and safety legislation (which is classified as an act of indiscipline), and appropriation of work center property for personal use continued to take up a lot of Council time up until the 1980s.

The second major kind of case which the Councils heard were those which fell under the general rubric of workers' rights violations. Included here are disputes over a host of issues central to workers' day-to-day experience on the job such as wages, working conditions, transfers, temporary or permanent layoffs, vacations, and demotions. The Councils also assumed the Grievance Commissions' previous task of processing disability, retirement and death pensions and protecting and verifying individual workers' claims to these benefits. Between 1965 and the late 1970s cases involving violations of workers' rights probably constituted an increasing percentage of the Councils' work: only 11% of the Council cases going to appeal in 1969 dealt with workers' rights; in 1972–73 17% of the cases before the Councils were concerned with these matters; by the next year this figures was estimated to have reached 25% (Hernández and Mesa-Lago, 1971, p. 223; Pérez-Stable, 1975, p. 73); three years later one province reported that at least 78% of the cases which went to the final level of appeal had to do with questions of workers' rights violations (*Trabajadores*, 3/22/77, p. 2). The growing percentage of workers' rights cases before the councils can be explained in part by the fact that they were now dealing with issues over which the earlier Grievance Commissions had no jurisdiction, such as wage disputes, evaluations and promotions. In addition, as Pérez-Stable suggests (1975, p. 73), the stronger emphasis after 1970 on tying remuneration to output probably increased the potential for wage and salary disputes between workers and managers.

Once open hearings have been held before the *Consejo*, its members issue their findings in accordance with relevant legislation. In disciplinary cases Councils may decide to uphold, change or nullify administration sanctions. Sanctions for indiscipline have varied somewhat over the years. Thus, during the so-called Sino-Guevarist period of the late 1960s, the *Consejos* were urged to stay away from sanctions of a material nature. However, along with the shift away from moral incentives in the 1970s, union leaders began to complain that the sanctions available to the Councils were too limited (Bray and Harding, 1974, p. 669), and throughout the 1970s the grievance bodies made determinations involving a wide range of punishments, from warnings, to permanent or temporary transfers with attendant wage adjustments, to dismissals. In cases where a Council found a violation of workers' rights, it could order that workers be reinstated in their jobs, granted wage adjustments, transferred to other worksites or other jobs, and paid back wages which had been partially or totally withheld. The Councils were also *expected* to notify appropriate judicial or administrative authorities when, in the course of deciding a case, they discovered or suspected that management-level personnel had acted improperly or illegally.[4]

If either a worker or an administrator were not satisfied with a *Consejo*'s decision, the verdict could be appealed at two higher levels: the Regional Appeal Council and the National Review Council. However, neither of these was composed solely of workers. The Regional Appeal Council was made up of three people: one representative each from the Ministry of Labor, management, and the union. The National Review Council was a five-member body, with two members each from the Ministry and management, and one from the CTC. Thus once a case was taken above the worksite level, its resolution was no longer in the hands of workers: the single union representative on either appeal body would have been unlikely to prevail when faced with the determined opposition of the Ministry and management representatives. Moreover, before 1970 the unions themselves were closely identified with the administration and the state bureaucracy, and so the balance of power on the two appeals commissions would have been even more skewed against workers than it appeared at first glance.

Workers' control over discipline and the enforcement of workers' rights was further restricted by the broad powers retained by the Ministry of Labor over the entire arbitration system. The Ministry, for example, could dismiss or change any member of a Work Council, and it could take any case in the process of being heard at any level and issue a verdict which was not appealable. It could also overturn a decision of any labor justice body, also leaving the involved parties without recourse to appeal (Hernández and Mesa-Lago, 1971, p. 221; Bonachea and Valdés, 1972, pp. 370–371; Zimbalist, 1975, p. 47). I have no information on how often the Ministry of Labor actually exercised its wide-ranging powers over the *Consejos* and appellate organs. However, their existence, even on paper, set important limits on workers' power to enforce legislation protecting their rights and to curb disciplinary abuses by management.

4. 1977: THE UNION TAKES OVER FOR THE MINISTRY OF LABOR; THE END OF THE TRIPARTITE APPEALS COMMISSIONS

Despite numerous changes in almost every major facet of the Cuban social system in the early and mid-1970s, the basic method of resolving workplace disputes which was put into place in 1965 survived until 1978. The preceding year, however, the National Assembly of People's Power (OPP) approved new legislation which spelled two important alterations in the system of Work Councils and appellate bodies. To begin with, full authority over the *Consejos* was transferred from the Ministry of Labor to the unions which did not, however, inherit the Ministry's earlier powers to override verdicts or dismiss Council members.[5] Thus, for the first time since the Revolution, the unions rather than an organ of the state administration oversaw the administration of labor justice. The shift was consistent with the overall trend of expanding union responsibilities and functions apparent in Cuba after 1970.[6]

The Law of the Organization and Functioning of the *Consejos del Trabajo* (Law No. 8) specified four different kinds of activities which the unions would undertake in association with their supervision of the Councils. The unions were to oversee all Council work, organize the election (and recall) of Council members, and ensure that members were prepared to carry out their duties. Last, the law mentioned a role for the union in the Council proceedings which took place at each worksite. Overseeing Council work included being available for consultation whenever Council members had questions, ensuring that worksite management provided the Councils with meeting space and necessary materials, and evaluating Council work by conducting on-site visits and arranging for *Consejo* members to report before meetings of the grassroots level union organization, among other things.

The procedure for selecting Council members over which the unions now took charge was much like the one which had existed since 1965: five members were elected by majority vote in all workplaces with over 25 workers. Qualifications for serving on the *Consejos* included: a good work record, adequate educational level, upstanding conduct outside work, and active incorporation into the Revolution. And though in most cases they would probably have had the necessary qualifications, both administrators and union leaders were specifically prohibited from being elected to a Council post. Workers I interviewed regularly mentioned other qualifications in answer to my queries about what kind of person made a good *Consejo* member: they must be dedicated and energetic because service on the Councils was a big responsibility and could take a lot of time; it was also very important that members be serious and objective people who would not, according to a member of the Cultural Workers' Union, "treat their friends better than their enemies."[7]

The general method used by the unions to train Council members was to include them in a variety of union-sponsored educational programs focused on labor legislation and workers' rights. Radio, television, and correspondence courses as well as self-study packets were prepared to this end. In addition, seminars, conferences and courses were sponsored both by the CTC and the various sectoral unions, sometimes lasting as long as six months. By mid-1978 the CTC reported that 40% of all Council members had received training in labor legislation, and plans were to increase this to 100% in the next year (*Trabajadores*, 10/14/78, p. 3). However, management at times was reportedly reluctant to facilitate workers' attendance at such training sessions, causing a certain amount of tension between workers and their bosses. To help alleviate this problem, the 14th National CTC Congress resolved to include arrangements for the ongoing education and training of Council members in the collective work commitments, the documents which specify the rights and obligations of both workers and managers at individual worksites throughout the country (*Trabajadores*, 12/2/78, p. 5).

The procedures for initiating, hearing, and deciding cases, as spelled out by the 1977 law, included a number of provisions designed to encourage workers' use of the grievance system: Work Councils continued to be composed solely of workers; lawyers were not required at any stage; proceedings could be initiated verbally; the law set strict time limits for each stage of the process, to encourage rapid resolution of cases; all Council hearings were public, and a notice listing the time and place of the hearings had to be posted at the worksite at least three days in advance; a worker could bring evidence or witnesses to support her or his position; workers could redress their grievances individually or in groups; they could choose to represent themselves, secure a lawyer or prevail upon anyone else to present their cases before the *Consejo*.

Nevertheless, in a situation where a worker

and a boss are at loggerheads — the defining feature of most Work Council cases since it was commonly felt that no case should be taken to the *Consejo* before efforts had been made to resolve it informally — these safeguards are not always sufficient to ensure that a worker utilizes the procedure when her or his rights have been violated, or when management has applied an unfair disciplinary sanction. What is necessary as well — and on this Cuban workers were adamant — is the organized, active and informed presence of the union and its membership to support, guide and advise workers embroiled in disputes with management at each stage of the conflict resolution process. "The union has got to be the one to watch that the *Consejo*'s actions are fair to the worker," one man told me.[8] Another worker stated that the union must serve as the "worker's advocate from the beginning of the conflict until it is finally resolved."[9] In essence, workers expected the union to take the place of the lawyer. Workers expected to be able to go to the local union before formal Council proceedings even began, to get help preparing their case and/or advice about how to respond to an administration allegation. Workers also hoped to get the union to argue on their behalf during the actual hearings (though a union officer can refuse to do this), and to ensure that the proceedings themselves were conducted properly and according to all the dictates laid down in the law. It was also explained to me that the union had an important role in making sure that any Council finding favorable to a worker was respected and enforced by the administration.

This kind of support was what workers I spoke with hoped to receive from their unions if they ever became involved in a dispute serious enough to be taken to the *Consejo del Trabajo*. However, conflicts between workers and management of this import are not everyday occurrences. A few workers I interviewed had never even heard of a dispute where they worked which made it to the Council. The majority had never been involved in one personally. Yet the following two stories, the first related to me by a cashier and the second by a worker in an assembly plant, illustrate how the unions' activities can make all the difference in the resolution of conflicts serious enough to demand Work Council action:

> When I think that both myself and another clerk were sent by the administrator to work somewhere else for three months just because of personality differences and a few pieces of badly marked merchandise! I cried when that happened to us, I don't mind saying. But in the end both of us were reinstated at our old jobs and it was the administrator who got booted down to an inferior one. What made the difference was our union. The union stuck with us from the beginning — arguing our case with the administrator's boss, at the *Consejo*, all the way up through our appeal. The union did it. They even packed the courtroom with our compañeros and union officials![10]

> I was the Secretary for Labor and Social Matters in my *sección* (the base-level union organization in Cuba). I was involved in trying to resolve absentee cases that came before the *Consejo*, and I began to notice that it was the same workers who were brought before the *Consejo* month after month for absenteeism. But I knew they were not the only violators in the factory. Many others were actually administrators. And so I brought this up with the administration. I said, "Listen, why don't you haul some of these administrators before the *Consejo del Trabajo* for absenteeism?" The administration was reluctant. I said, "I can prove that of the 24 administrators' time cards in this work center, 18 show unjustified absences!" But still the administration balked at bringing a case. I tried other routes. I talked to the union, but the union wouldn't back me up. It was just too hard for the union leaders to force the issue because it meant going against *jefes*. The situation was never resolved. The *Consejo* could never do anything at all.[11]

Along with the transfer of authority over the *Consejos* to the unions, a second major change occurred in the Cuban system of dispute resolution after 1977. This change involved appeals, which as we saw earlier, had previously been handled by commissions composed of management, Ministry of Labor and union representatives, who were in the minority. After 1977, however, appeals of Work Council verdicts were dealt with in the newly created labor chambers (*salas de lo laboral*) of the regular Cuban court system.[12] Thus the Councils came to represent the last opportunity to resolve a labor–management dispute before the matter spilled over into the formal legal arena. As a result, according to one informant, Council members tried very hard to get both parties in a dispute to agree to their decision.[13] However, if either side expressed disagreement with the *Consejo*'s verdict within 10 days, the decision was not binding, and the case could be appealed first to the Municipal and later, if necessary, to the Provincial Tribunal. Or, if the Council found in favor of a worker but the administration did not comply with the ruling, the worker had six months in which to initiate appeal proceedings (*Trabajadores*, 1/4/79, p. 3). Some of the same aspects of the *Consejo* procedures designed to encourage workers to make use of the process were also incorporated into the procedures of the labor chambers of the Municipal and Provincial Tribunals.

A major question to be considered with regard to the new appellate system is whether it represents a net increase or a decrease in workers' power in the resolution of labor–management conflicts. Information of two types would be helpful in trying to answer this question. The first would be data on the outcome of Work Council cases going to appeal both before and after 1977. Unfortunately, I have no such comparative information; we only know that in 1981, the CTC reported that the Municipal Tribunals upheld only 47% of the cases they heard in which workers had been fired (*Trabajadores*, 3/31/82, p. 2). We would also benefit from detailed information on the amount of influence workers have over the selection, and thus the composition, of the Tribunals. Appeals bodies operating before 1977 were far from ideal in terms of either worker representation or worker involvement in the process of selecting members. Currently, Tribunal judges who preside over appeals of Work Council cases are elected by the corresponding Assembly of People's Power. An indirect result of this method of selecting judges could conceivably be an expansion of working people's influence over the resolution of labor disputes brought to appeal.[14] The judges, however, are not nominated by the Assemblies themselves, but are elected from a slate of candidates submitted by the Ministry of Justice. Yet these slates are compiled on the basis of proposals from the unions, the party, and certain relevant state agencies.[15] However, I have too little information on the details of how the unions decide on their nominee preferences or how much weight union recommendations have in the final compilation of the candidate slates, to know whether the new judicial selection system is open enough to result in benches which genuinely reflect the preferences of the mass of Cuban workers.

The types of cases which fell under the purview of the Councils after 1977 were similar to those which had concerned them previously. On the basis of what workers told me and reports in the union press, absenteeism continued to account for a large percentage of the indiscipline cases heard by the Councils. In fact Roberto Veiga, Secretary-General of the CTC, reported in early 1980 that absenteeism and tardiness accounted for 85% of such cases which came before the *Consejos* (1980, pp. 30–31). Other kinds of indiscipline cases commonly heard by the Councils included taking worksite property for personal use, damaging or wasting property or materials, disrespect or abuse of a fellow worker or superior, and negligence or carelessness in production or service delivery.[16] Workers indicated to me that Work Councils were loath to uphold the firing of a fellow worker, but that this definitely occurred on occasion, particularly in cases of repeated and serious indiscipline. Lone yet serious infractions of work discipline most often incurred penalties such as temporary transfers to lower paying jobs or postponement of vacations. It is important to remember that during this period Work Councils *had* to be involved any time an administrator applied a sanction at the workplace, regardless of the seriousness of the alleged offense. In cases of very serious indiscipline or criminal activity, an administrator could sanction a worker at once. However, the following day the administration was required to go before the Work Council, which would review the case and uphold, revoke, or modify the sanction (*Trabajadores*, 7/5/77, p. 1; 12/19/78, p. 2). In less serious cases, the administration could not apply a sanction at all without first bringing the case before the Council, which then decided whether circumstances warranted the requested sanction.[17] Administrators who violated these procedures were subject to imprisonment, a fine, or both.[18]

The *Consejos* also continued to review cases of workers' rights violation after 1977.[19] Again based on what workers reported to me and on articles in the union press, one of the most common disputes of this sort involved remuneration.[20] It is not hard to understand why this issue in particular might spark controversy in Cuban enterprises. A worker's basic wage or salary is based on a detailed scale containing numerous classifications, and both the scale and the job classifications are subject to periodic revisions. Moreover, how much any worker ends up making is also affected by a variety of other regulations governing historical wages,[21] difficult or unsafe working conditions, sickness and accidents, and interruptions in production. The determination of a worker's wages for any given period can therefore be a complex affair, lending itself to a variety of possible interpretations. In addition, efforts in the 1970s to strengthen the link between pay and output in as many occupations as possible, and the post-SDPE stress on enterprise "profitability," have combined to make remuneration a prime focus of worker–management disputes.

A second very common type of workers' rights case involved transfers.[22] A 1969 regulation governing transfers left a great deal of room for disagreement between workers and state administrators. According to this regulation, individual workers had the right to change worksites as well as the kind of work they do. Yet that right had to be balanced by society's need to distribute the labor force rationally (*Trabajadores*, 11/7/78,

p. 3). Administrators had 60 days in which to respond to a worker's request for a transfer, and a negative response could be contested before the Work Council. The Council then had the unquestionably difficult task of balancing the administration's case for denying the request against the worker's desires to be closer to ailing family members, to continue her or his studies, to accept a better paying post, and so on. However, with the increased use of employment contracts, which can be terminated by the worker at will so long as the proper amount of advance notice is given, this type of case may become less common.

Despite its strengths in terms of involving direct producers in the process of settling major workplace disputes, the post-1977 grievance system I have just reviewed has been criticized by management, union officials and workers alike. The kinds of problems I learned about include some which might be overcome through adjustments or improvements in the system itself. For instance, some *Consejo* members were said to lack the necessary knowledge of labor and social welfare legislation; people complained that it took far too long to settle cases; some workers lost cases simply because they were unaware of time limits for bringing them before the *Consejos* or appeal bodies; some Councils were criticized for being too lenient, others for being too strict. "We have problems with *Consejos* of both stripes," one worker remarked.[23] Solutions to such problems seem fairly straightforward. The union could expend additional energy training and educating *Consejo* members as well as ordinary workers. More *Consejos* could be formed. And workers themselves could elect stricter (or more lenient) Council members in workplaces where they were dissatisfied with the way in which the grievance resolution organs functioned.

However, there were other difficulties with the conflict resolution mechanisms which cannot be rectified quite so simply, because they are rooted in continuing power differentials between Cuban workers and managers. These kinds of problems, which result from management misuse or abuse of the Councils, will demand more sustained and ingenious corrective efforts on the part of the workers and their unions. The union press has published a number of reports which point out ways in which state managers have been able to thwart workers' ability to control the outcome of disputes over discipline and workers' rights through the Councils. Administrators, for example, have simply failed to appear at Council hearings after a worker has filed a complaint against them (*Trabajadores*, 5/15/79, p. 3). Or a management representative who is not involved and knows little about the case at hand has been sent to testify at the proceedings (*Trabajadores*, 11/7/78, p. 3). Administrators have also been criticized for all sorts of improprieties in their handling of disciplinary cases. According to one worker, managers conveniently missed deadlines for presenting disciplinary cases to the Councils when certain workers were involved, thus invalidating the sanction they had "imposed."[24] Some administrators asked for very mild sanctions for relatively serious offenses, while others hauled workers with exemplary records before the Councils for very minor offenses, or ones for which the worker turned out to have a reasonable and just explanation. Other managers let minor instances of indiscipline accumulate until a worker could be charged with a relatively major offense.[25] Despite the risk of the penalties they might incur, managers were known to bypass the Work Councils entirely, applying sanctions without notifying the workplace grievance body, hoping that the Council would prefer to ignore the matter once it finally got wind of it.[26] The most common complaint of all, however, was that some managers simply ignore the rulings of the Work Councils (and appellate bodies as well at times), refusing to modify or suspend disciplinary measures they had imposed, or declining to make reparations in cases where they have been found to have infringed on the rights of a worker.[27] In cases such as these much depends on the union. Where it is weak and deferent, managers will be far more likely to succeed in attempts to short-circuit the Work Councils. Despite the Council structure, the power to discipline and to interpret workers' rights legislation would effectively remain in management's hands. But where the union is alert and combative, and enjoys the active support of the rank and file, managers will find it much more difficult to circumvent the legal proscriptions governing the resolution of workplace conflicts through the *Consejos* and appellate bodies.

5. LAW NO. 32: THE *CONSEJOS* LOSE THEIR AUTHORITY OVER WORKPLACE DISCIPLINE

At the end of February 1980, the Council of State passed a law which most workers I spoke with simply referred to by its number — 32. Law No. 32 substantially altered the functions of the *Consejos del Trabajo* by abrogating their jurisdiction over cases of indiscipline at the workplace. Managers in Cuba can now discipline workers without notifying or involving the Work Council at all.[28] Law No. 32 requires only that sanctions be imposed within 30 days of the act of indiscipline, that the worker affected be notified in writing, that the union receive a copy of this document, and that the administrator take into account the gravity, consequences and circumstances surrounding the act of indiscipline, along with the worker's past conduct. The first recourse of a worker who feels a disciplinary measure has been applied unjustly is now the labor chamber of the Municipal Tribunal.[29] Disciplinary measures imposed by management remain in force *until* the Municipal Tribunal (or the Provincial Tribunal in the case of a further appeal) renders a verdict, whereas previously sanctions could be enforced only *if* approved by the *Consejo* or *after* a worker had received an unfavorable ruling from the Municipal (or the Provincial) Tribunal. The *Consejos del Trabajo* continue to operate as they had previously in cases of alleged violations of workers' rights — worker–management disputes over wages and salaries, vacations, transfers, promotions, etc. — matters of importance to workers everywhere.[30] The unions have not abandoned their involvement with the Councils, although in 1982 they considered decreasing their number.[31] Since Law No. 32 obviously constitutes a major departure from the earlier trend toward increasing worker control over workplace dispute resolution, we must examine both the reasons for this reversal and what Cuban unions and Cuban workers think about it.

The genesis of Law No. 32 can be traced to a growing sense among Cuba's political leadership in the late 1970s that workplace indiscipline was reaching unacceptable proportions. Castro made comments to this effect before the National Assembly of People's Power in 1979, and in October of that year a study of labor discipline by commissions which included local OPP and union representatives was begun. The plan was to visit 20% of Cuba's workplaces and gather information from workers, administrators, union officers, *Consejos del Trabajo*, as well as from Municipal and Provincial Tribunals. A preliminary report was presented to the National Assembly in December 1979, after half the scheduled work centers had been visited. It called attention to a variety of problems including deficiencies in existing labor legislation and in the functioning of the *Consejos del Trabajo*; lack of knowledge about labor legislation on the part of administrators, *Consejo* members, and unions; lackadaisical and/or overly mechanical use of grievance procedures by administrators; and the amount of time it was taking for the Councils and appeals bodies to decide on cases.[32]

This last point was emphasized by a number of workers in response to my own queries about why Law No. 32 was promulgated. The Work Councils, to summarize what some of them told me, just couldn't handle all the cases before them. As one worker recalled, "The *Consejos* had a hand in everything. They had way too much to do!"[33] Or as another remarked, "The *Consejos* were causing divorces, I swear! Members had to meet for hours and hours."[34] Workers' perceptions are corroborated by other evidence. As the 1970s progressed more cases were being brought to the *Consejos*: 161 cases were heard by the Councils for every 10,000 Cuban workers in 1972; in 1974 the analogous figure was 270; in 1976 it was 280; in 1978 the figure had reached 306. Moreover, a common complaint in the late 1970s concerned the Councils' backlog of unsettled cases and the excessive amount of time they were taking to resolve workplace conflicts. There were reports in the union press during the period of cases taking one and a half, three, and even four years to settle (*Trabajadores*, 1/25/77, p. 4; 7/8/77, p. 4; 11/7/78, p. 3; 3/17/79, p. 3; 6/5/79, p. 3 and 10/13/79, p. 5).

However, even though the *Consejo* system was getting bogged down by the late 1970s, this can be considered the "cause" of Law No. 32 only in a secondary sense. More correctly, heavier use of the Councils should be considered the symptom of two significant changes which occurred in the 1970s: union revitalization and the institution of a new management and planning system, the SDPE.[35] Quite likely these two developments, in combination, created pressures which spearheaded the promulgation of Law No. 32. Stronger and more aggressive unions, as I hinted earlier, served to encourage workers to challenge administrators when they felt they had been wronged, and to discourage managers from instituting disciplinary measures without properly involving the Councils. At the same time, because a primary aim of the SDPE was to induce enterprises to operate more efficiently and to cover their costs with their income and turn a "profit" besides, it served both to focus management attention on the maintenance of high levels of discipline at the workplace, and to increase conflict over workers' rights insofar as managers responded to their new operating environment by ignoring (or bending) laws governing remuneration, health and safety, or rest periods, for instance. And at the same time that both the changing unions and the new management and planning system were placing new demands on the Council system, the compelling realities of Cuba's economic difficulties as a small developing country heavily dependent on one export crop in a period of spiraling world inflation, were increasing the pressure on the Councils to deal with cases more swiftly, more efficiently, and even more strictly than they were prepared, or perhaps even inclined, to do.[36] The *Consejos*, caught in the middle, were then relieved of their authority over workplace discipline by Law No. 32.

The tension in contemporary Cuba between what Pérez-Stable has termed "unionist" (*obrerista*) and the "rationalist" (*eficientista*) tendencies was resolved in this instance in favor of the "rationalizers" (Pérez-Stable, 1983, p.7). The danger that, despite increased union strength, a minority strata of "rationalizers" will go on to acquire more workplace power should not be brushed aside lightly. After all, with Law No. 32 an impressive and longstanding revolutionary tradition which granted Cuban workers a great deal of control over workplace discipline was nullified by a single decree. It is not hard to imagine how, by removing the arbitration of disputes over discipline from the shop floor and from the hands of fellow workers, Law No. 32 might make it more complicated, more intimidating, more time consuming, and less profitable for workers to counter unjust disciplinary measures imposed by administrators. In light of this, it is important to look closely at the reactions of the unions and of individual Cuban workers to the institution of "*Treinta y dos.*"

For their part the Cuban unions lent their quick approval to Law No. 32. At the same time, however, the law sparked more intense debate and discussion in the workers' organizations during the first half of the 1980s than any other labor–management issue. The principal point of controversy concerned management abuse of its newly gained disciplinary powers. Specifically, mangers were frequently charged with overzealous use of the most severe sanctions possible under the law, and with applying sanctions without taking circumstances, consequences or the worker's past conduct into account: the union announced that the Tribunals were upholding the implementation of the maximum sanction (firing) in only 47% of the instances in which the worker appealed her/his punishment; extenuating circumstances were found to have been ignored by management when applying Law No. 32 in nearly 40% of the cases reviewed.[37] Indeed, two workers I interviewed felt this is precisely what had occurred in their cases:

> It hurt me a lot when 32 was used against me. Someone in our office had been ill and basically I was doing two people's jobs, and as a consequence I miscounted some money. Even though I only drew a warning, I don't think it's right at all.[38]

> I sure wasn't happy when they used 32 against me. I knew someone was mismarking a lot of merchandise but I didn't want to rat on them. But when it was discovered, the *jefe* said I was to blame too, because I knew about it. So 32 was applied, though I wasn't sanctioned severely. But I have a good work record, so I ask you, was that really fair?[39]

What turned out to be especially problematic from the unions' point of view was the unequal vigor with which Law No. 32 and a second law concerned with managerial discipline were being used. Law No. 36, as this latter law is called, was promulgated at the same time as Law No. 32, and although the list of possible infractions contained in the law was quite long, it was to be applied by administrators as opposed to the *Consejo* and it was appealable at the Ministerial, not the work center level.[40] Although one worker's guess that application of Law No. 32 had been "forty times higher than 36" was off, figures from a CTC survey in the first months of 1981 indicated Law No. 32 was being used nearly 25 times more often than was Law No. 36, a figure felt excessive even in light of the administrator to worker ratio in the labor force.[41]

In view of the dissatisfaction surrounding the implementation of Law No. 32, union leaders have repeatedly called for union vigilance against abuses of the law and for more rigorous use of Law No. 36 against administrators who either incorrectly apply disciplinary sanctions or who violate workers' rights. In addition, at the 15th CTC Congress in 1984 the need for legal mechanisms to rehabilitate the work records of those who have been sanctioned under Law No. 32 was expressed (Veiga, 1984, p. 30). Finally during discussions of the Theses of the 15th Congress held in work centers all over the country months before the event, many union members argued that administrators who applied disciplinary sanctions which were later overturned by a Tribunal should compensate the wronged workers out of her or his own salary. This idea came up again at the Congress itself, though many participants were opposed to it, arguing that it would make managers reluctant ever to apply Law No. 32. As far as I know unjustly sanctioned workers continue to be compensated from the budget of the enterprise.

Individual workers I spoke with invariably brought up the topic of Law No 32, and generally said they saw a need for it. They made a connection between the quantity and quality of goods and services in Cuba and how workers performed at their jobs, and so welcomed a tighter and less cumbersome system than the *Consejos* had previously provided for dealing with people who didn't show up to work, damaged or stole work center property, or neglected health and safety regulations, for example. "There really were some workers who weren't working before and the administration couldn't do a thing about it," according to one worker.[42] "Everyone else had a law protecting them," said another. "Couldn't we just have one little law for the administrators?"[43] However, despite their general approval of the law, worker support for 32 was not unqualified. Most recognized the potential for management abuse which stemmed from the power they gained under the new law:

> Law No. 32 will be a good law, if it is applied properly. It gives more power to the administration. Before they did not have enough. The law itself is just. The problem could come with its application. There is a danger that with this increased power bosses could use the law to get at workers they don't like or to invent charges against workers. But we had a lot of problems with discipline before. Workers with consciousness support the law. Workers without consciousness don't support Law No. 32, but then they didn't support the Work Councils dealing with discipline either.[44]

Such wariness, both on the part of the unions and the individual workers who comprise them, will hopefully moderate Law No. 32's negative effects on the worker–management balance of power, against the day when Cuban workers might once again assume the major share of responsibility for discipline at their worksites.

6. CONCLUSION: THE IMPORTANCE OF UNIONS FOR CUBAN WORKERS' POWER

According to one student of power at the workplace, a dispute resolution body independent of management and preferably composed of peers, though empirically rare, is one "minimally necessary" component of workplace democratization (Bernstein, 1976, especially Chap. 8). I would agree with Bernstein's basic assessment and would therefore argue that the development of grievance mechanisms in post-revolutionary Cuba — especially the *Consejos del Trabajo* — offers much of interest to anyone concerned with equalizing the division and exercise of power in the course of production.

However, this investigation of how conflicts have been resolved at the Cuban workplace, underscores an important point about power at work which deserves more attention. No matter how egalitarian the *form* of any workplace decision-making structure — and the Cuban *Consejos* embody much worth emulating in this regard — so long as the minority of managers possess greater amounts of skill, education, confidence, time, status, money, etc. than the majority of workers, it is almost certain they will be able to employ these advantages so as to prejudice the workings of these structures in their favor. The division of power, in other words, will remain badly skewed despite democratic appearances. Workers may devise various solutions to this problem but one of the most important must be organization. Individual workers must make up for some of the advantages they lack in other areas through unionization. They will have a better chance of translating the formal *right* to decide a particular issue into the *power* to ensure that the issue is actually decided in accordance with their own preferences, if they are supported by an organization which unifies their class.

Briefly, I will review some of the ways in which unions have proven central to worker control of the grievance resolution mechanism in Cuba. Before doing this, however, I must anticipate an important objection: involvement of unions in grievance resolution, or certainly any other decision-making arena, does not guarantee greater worker power. The union itself may exist but be too weak in terms of the percentage of the workforce organized or the balance of power at the national level to effectively counter management advantage. Or somewhat ironically, as we saw in the Cuban case, active unions can create pressures which result in a decline in workers' power in certain decision-making areas. Finally, for a variety of reasons, the union might more correctly be considered a supporter of the status quo at the workplace, rather than an advocate of greater assumption of power by the workforce. Under capitalism this might occur in the case of company unions or when unions become a pillar of what Burawoy has termed the internal state — institutions within the firm's jurisdiction which regulate conflicts over production relations (Burawoy, 1979, especially Chap. 6). Similarly, some observers of socialist countries have argued that the unions there are only transmission belts of state and the party directives and that therefore, as Hazard (1968) has written with regard to the Soviet Union, "the transfer of the grievance procedure to the trade-union organization is not the concession to labor that it appears to be."[45]

Interestingly, in socialist Cuba, in part because of the close identification of prerevolutionary unions with the Batista government, union officers have always been prohibited from serving on the *Consejos*, and the union has no obligatory role in their proceedings. Nonetheless, as I have suggested at various points, unions have been vital in making sure workers are able to use the opportunity provided by the dispute resolution structures to enhance their power at the workplace. In a number of ways Cuban unions, despite purposeful limits on their formal involvement, have prompted grievance resolution mechanisms to correct, rather than only reflect, power inequalities at the workplace.

To begin with, the union's presence has provided workers who feel management has violated their rights with the security and moral boost necessary to initiate proceedings against their *jefes*. The union has played an analogous role in the often touchier situation when a worker has been disciplined unjustly. Here the union's backing has helped workers overcome their understandable reluctance to challenge a boss' sanction. And in both kinds of cases the union — through educational activities among its members — has increased workers' awareness of precisely what their rights are and what constitutes an act of indiscipline, thus making them more likely to recognize management infringements of these rights or inappropriate applications of disciplinary sanctions.

Unions have also helped workers secure favorable resolutions of their disputes with managers by counseling them on the preparation of their cases, making sure that procedures are followed correctly by the *Consejos* and the managers, and by arguing workers' cases before the grievance bodies. Moreover, as we saw earlier, the union has tried to make sure that managers abide by the verdict when *Consejos* or appeal bodies find in a worker's favor, either correcting their earlier action and/or compensating the wronged worker as ordered. It is also important to recognize that the independence of the five worker-members of the *Consejos* — vis-à-vis management — is heightened insofar as the overall balance of power at the worksite is equalized through the presence and activities of the union. Last of all, union backing in many of these ways has been especially important whenever a worker's case proceeds to the appeal level, where procedures are more complicated and formal and where cases are not heard by a panel composed of one's peers. For all these reasons appeals can prove especially intimidating to an untutored worker, in the absence of the collective support of other workers through the union.

Although the data are not complete, figures on the use of *Consejos* and appeal bodies, when considered in conjunction with changes that have occurred in the unions, do provide some confirmation for my argument concerning the importance of unions for the division and exercise of power *vis-à-vis* conflict resolution in Cuba. Briefly, after the early 1970s, Cuban unions were much stronger and more active organizations than they were during the first decade of the Revolution. Their functions expanded considerably; they carried them out in a more autonomous fashion; they organized a greater percentage of the labor force; they were financed by members' dues rather than out of state coffers; leaders at all levels began to be selected according to regularized and frequent electoral procedures (Fuller, 1985, Chaps. 4, 5, 10). We should therefore expect grievance mechanisms to be more heavily utilized after 1970, and indeed we do find some evidence to suggest this. Between 1967 and 1969, appeals of *Consejo* cases heard by the Regional Appeal Council and the National Review Board dropped. Despite an increase in the labor force of around 119,000 persons, the latter organ heard approximately half as many cases in 1969 as in 1967, and the former body heard only around two-thirds the number of cases in 1969 as it did two years previously. In contrast, though the grievance procedure itself did not change, between 1972 and 1978 the number of cases heard by the *Consejos* more than doubled; between 1969 and 1976 the number of cases before the Regional Appeal Council increased nearly sevenfold, and in the same time period the number before the National Review Council increased almost elevenfold — all three figures far outpacing the growth in the labor force. As was to be expected, after the 1980 passage of Law No. 32, the number of cases before the *Consejos* dropped considerably because they now heard only workers' rights cases. However, since there is no evidence that the unions became weaker after 1980, we should expect no drop in appeals cases, and indeed they appear to have increased. I estimate that almost 85,000 cases per year reached appeal between 1980 and 1983, almost three times the number in 1976.[46]

The foregoing analysis of dispute resolution suggests that unions are a critical factor in the consideration of power in postrevolutionary social relationships at the Cuban workplace. During the period when unions were the smallest, the weakest, and the most dependent on the state administration and the party, the grievance mechanisms, though exemplary in many regards, were least able to remedy the effect of unequal power divisions on those workers who had experienced them in a firsthand and overt fashion. Yet after 1970, as unions became stronger, larger, and more active, they have served to encourage individual producers (who on their own may not have tried or chosen to do so) to challenge their boss's power through the existing grievance resolution mechanisms. And since 1980 when the method for settling disputes over discipline was altered in favor of management, the unions' role in uniting individual Cuban producers into a collective force capable of extending workers' power at the point of production has remained critical, not only in situations of overt workplace conflict, but at other times as well.

NOTES

1. For these differing accounts, see Zeitlin (1970, pp. 191–192) and CERP (1963, p. 134).

2. CERP (1963), p. 134. However, Zeitlin's characterization of what kinds of cases the *comisiones* could hear is somewhat different. See Zeitlin (1970), p. 192.

3. Interviews with a member of the Education and Science Workers' Union, Havana, 5/9/82 and 9/23/82.

4. See for example, *Trabajadores*, 1/25/77, p. 4; 3/18/77, p. 2; 9/27/77, p. 2; 9/30/77, p. 2; 10/14/77, p. 2 and 8/26/77, p. 2.

5. This change was contained in "Ley No. 8 de la Organización y Funcionamiento de los Consejos del Trabajo" (Law No. 8). See especially artículo 10. As part of a government reorganization, the Ministry of Labor was dissolved around the same time and some of its activities were taken over by the new State Committee of Labor and Social Security (CETSS).

6. See Fuller (1985), Chap. 4.

7. Interview, Havana, 9/25/82.

8. Interview with a member of the Health Workers' Union, Havana, 9/27/82.

9. Interview with a member of the Cultural Workers' Union, Havana, 9/25/82.

10. Interview with a member of the Food and Commercial Workers' Union, Havana, 10/5/82.

11. Interview with a member of the Light Industry Workers' Union, Havana, 10/9/83.

12. Around this time the Cuban judicial system underwent a number of other important changes: the courts were reorganized in accordance with the new politico-administrative division of the island; the earlier all-lay Popular Tribunals (*Tribunales Populares de Base*) were replaced by the Municipal Tribunals (*Tribunales Municipales*); the entire court system was unified; and new legal codes were promulgated in all areas of law, replacing some in force since before the revolution. See Berman and Whiting Jr. (1980, pp. 478–479); Domínguez (1978, p. 257); and Alvarez (1981, pp. 363–395) for more details.

13. Interview with a member of the Cultural Workers' Union, Havana, 9/25/82.

14. The Cuban judicial system allows for both professional and lay judges on the Tribunals at all levels. On the Municipal and Provincial benches, the lay judges outnumber the professionals. Only the Assembly which elected a given judge has the power to recall her or him, and each judicial body must report on its activities before the Assembly which elected it at least once a year. See Álvarez (1981), pp. 390–391.

15. During their tenure Cuban judges cannot simultaneously hold administrative or executive positions or jobs. Apparently this restriction applies to both lay and professional judges. See Álvarez (1981), p. 376.

16. A member of the Education and Science Workers' Union told me that, though it occurred only infrequently, teachers had been brought before the Work Council in his sector for engaging in amorous relations with their students. The Ministry of Education has rules against this kind of conduct, he explained (interview, Havana, 7/12/83).

17. The administration had 15 to 90 days to bring the case to the Council, depending on the gravity of the alleged violation. In instances of more serious violations, more time could elapse before the administration notified the Work Council that it intended to sanction a worker. See Ley No. 8, artículo 17.

18. Violations of this type could incur fines of up to 500 *cuotas* and prison terms of up to three years; see *Trabajadores*, 12/19/78, p. 3; 12/23/78, p. 2; 11/10/79, p. 2.

19. A worker had 180 days in which to bring a case of a violation of her or his rights before the Work Council; see Ley No. 8, artículo 18.

20. For details of some cases of this kind see *Trabajadores*, 11/16/78, p. 3; 2/20/79, p. 3; 2/22/79, p. 3; and 8/23/79, p. 3.

21. Some workers receive a higher wage than others for doing the same job simply because they held the job at some particular historical period, usually before the Revolution, when wages for that job were above the current rate. This is what is meant by the term "historical wage."

22. For a few examples of this type of case see *Trabajadores*, 11/7/78, p. 3; 1/4/79, p. 3 and 6/5/79, p 3.

23. Interview with a member of the Education and Science Workers' Union, 7/12/82.

24. Interview with a member of the Public Administration Workers' Union, Havana, 9/29/83.

25. See, for example, Veiga (1980), pp. 30–31 and *Trababjadores*, 10/6/79, p. 1; 2/24/79, p. 4 and 8/25/79, p. 4.

26. See, for example, *Trabajadores*, 8/17/78, p. 2; 12/19/78, p. 3.

27. See, for example, *Trabajadores*, 8/23/79, p. 3; 5/29/79, p. 3; 1/4/79, p. 3; 7/8/77, p. 4; 3/22/77, p. 2 and 3/19/76, p. 2; Veiga (1978), p. 25.

28. The acts which constitute violations of workplace discipline are discussed in *Trabajadores*, 11/24/81, p. 3; 11/25/81, p. 3.

29. This must be done within 10 days and in writing, though in a very simple format; see Decreto-Ley artículo 5.

30. After 1980 retirement pensions were no longer handled by the *Consejos* either (CTC, n.d., p. 13).

31. *Trabajadores*, 3/15/82, p. 2. See also 4/1/81, p. 2. It was reported at the 15th CTC Congress in 1984 that between 1979 and 1983, 74% of the findings issued by the *Consejos* were accepted by both parties involved (CTC, 1984, p. 148).

32. The final report with recommendations for corrective action was to be presented to the National Assembly the following year, but by then Law No. 32 had been promulgated. The preliminary report was published in *Trabajadores*, 12/29/79, p. 4. See *Trabajadores*, 3/15/80, p. 2 for additional details on some of the perceived difficulties with the dispute resolution system.

33. Interview with a member of the Cultural Workers' Union, Havana, 9/25/82.

34. Interview with a member of the Cultural Workers' Union, Havana, 10/6/83.

35. On these two developments see Fuller (1985), Chaps. 4, 5, 7, 10.

36. Law No. 32 was seen not only as a remedy for inadequately functioning Councils *per se*, but also as a corrective for administrations which failed to utilize the Councils as fully or aggressively as they might. Of course, if they had done this, it would only have further increased the Council's business, which I have argued was an indirect cause of the promulgation of Law No. 32 in the first place. See for example *Trabajadores*, 2/26/80, p. 1; 2/21/80, p. 3.

37. *Trabajadores*, 3/31/81, p. 2; 6/17/81, p. 2; For other examples of criticisms of Law No. 32 see 6/1/81, p. 1; 3/27/82, p. 3; 3/14/81, p. 3; 3/15/80, p. 2.

38. Interview with a member of the Public Administration Workers' Union, Havana, 10/13/83.

39. Interview with a member of the Food and Commercial Workers' Union, Havana, 10/12/83.

40. See Pérez-Stable (1983), p. 7 and *Trabajadores*, 3/23/82, p. 3. See *Trabajadores*, 3/9/82, p. 3 and 3/11/82, p. 3 for a listing of what constituted an infraction of administrative discipline under Law No. 36.

41. Interview with a member of the Public Health Workers' Union, Havana, 9/25/83 and *Trabajadores*, 6/17/81, p. 2; 9/15/81, p. 3. See also Veiga's comments on the imbalance 3/14/81, p. 3.

42. Interview with a member of the Public Health Workers' Union, Havana, 9/25/83.

43. Interview with a member of the Agricultural Workers' Union, Havana Province, 10/8/83.

44. Interview with a member of the Education and Science Workers' Union, 7/12/83.

45. Hazard (1968). See also Markovits (1982) for a discussion of the unions and the workplace conflict commissions in the GDR.

46. Data on numbers of cases before the *Consejos* and appeal bodies are scattered and incomplete, and so my calculations should be considered best guesstimates. See Fuller (1985), p. 365; *Trabajadores*, 6/17/81, p. 2; 3/15/82, p. 2; CTC (1984), p. 148; Veiga (1984), p. 32.

REFERENCES

Álvarez-Tabío Fernando, *Comentarios a la constitución socialista* (La Habana: Editorial de Ciencias Sociales, 1981).

Berman, Harold, and Van Whiting, Jr., "Impressions of Cuban Law," *American Journal of Comparative Law*, Vol. 28 (Summer 1980), pp. 475–486.

Bernstein, Paul, *Workplace Democratization: Its Internal Dynamics* (Kent, Ohio: Kent State University Press, 1976).

Bonachea, Rolando, and Nelson Valdés, "Labor and Revolution: Introduction," in Rolando Bonachea and Nelson Valdés (Eds.), *Cuba in Revolution* (Garden City, N.Y.: Anchor Books, 1972), pp. 357–383.

Bray, Donald, and Timothy Harding, "Cuba," in Ronald Chilcote and Joel Edelstein (Eds.), *Latin America: The Struggle with Dependency and Beyond* (New York: John Wiley and Sons, 1974), pp. 583–734.

Burawoy, Michael, *Manufacturing Consent: Changes in the Labor Process under Monopoly Capitalism* (Chicago: University of Chicago Press, 1979).

Central de Trabajadores de Cuba (CTC) *XV congreso de la CTC memorias* (La Habana: Editorial de Ciencias Sociales, 1984).

CTC, "XV congreso de la CTC proyecto de tesis," n.d., pp. 1–23.

Cuban Economic Research Project (CERP), *Labor*

Conditions in Communist Cuba (Miami: University of Miami, 1963).

Domínguez, Jorge, *Cuba: Order and Revolution* (Cambridge: Belknap, 1978).

Fuller, Linda, "The Politics of Workers' Control in Cuba, 1959–1983: The Work Center and the National Arena," PhD dissertation (Berkeley: University of California, 1985).

Hazard, John, *The Soviet System of Government* (Chicago: University of Chicago Press, 1968).

Hernández, Roberto, and Carmelo Mesa-Lago, "Labor organization and wages," in Carmelo Mesa-Lago (Ed.), *Revolutionary Change in Cuba* (Pittsburgh: University of Pittsburgh Press, 1971).

Markovits, Inga, "Law or order — Constitutionalism and legality in Eastern Europe," *Stanford Law Review*, Vol. 34 (1982), pp. 513–613.

Pérez-Stable, Marifeli, "Whither the Cuban working class?" *Latin American Perspectives*, Vol. 2, No. 4 (1975), supplement, pp. 60–77.

Pérez-Stable, Marifeli, "Cuba en los 80," *Areito*, Vol. 8, No. 32 (1983), pp. 4–9.

Trabajadores. La Habana.

Veiga, Roberto, "Informe central presentado al XIV Congreso Nacional de la CTC," *Trabajadores*, special supplement, 11/29/78, pp. 1–29.

Veiga, Roberto, "Clausura del octavo curso para directores de empresas, celebrada en la Escuela Nacional de Dirección de la Economía," *Cuestiones de la economía planificada*, Ano 3, No. 2 (marzo–abril 1980), pp. 24–34.

Veiga, Roberto, "Report Presented to the 15th Congress of the CTC" (1984).

Zeitlin, Maurice, *Revolutionary Politics and the Cuban Working Class* (New York: Harper and Row, 1970).

Zimbalist, Andrew, "Worker participation in Cuba," *Challenge*, Vol. 18, No. 5 (November-December 1975), pp. 45-54.

·10·

Cuban Planning in the Mid-1980s: Centralization, Decentralization, and Participation

GORDON WHITE

1. INTRODUCTION

Imagine parenting two sons: a four-year old with a six-year old. The four-year old has taken to hiding and losing puzzle pieces. Further suppose that your younger son has a friend who comes to visit and paints over several of the remaining pieces, creating a different image. Completing such a puzzle is a bit like analyzing Cuba's system of economic planning. (Zimbalist, 1985, p. 213)

This paper seeks to describe and analyze the current system of economic planning in Cuba in the light of contemporary reform currents in other state socialist economies, both industrialized and industrializing. The past two decades have brought widespread recognition of the economic deficiencies of centralized directive planning systems in Eastern Europe and economic reform programs have emerged which differ in nature (for example, contrast Hungary and the GDR) and in scope (contrast Hungary and the Soviet Union) (Nove, 1983). More recently, the reform impetus has emerged in Third World socialist countries, notably China since 1978 (Feuchtwang and Hussain, 1983) and, to a lesser degree, Vietnam since 1979 (White, 1983).

There have been three complementary elements common to most of these attempts to reform traditional planning systems: a shift from directive to parametric or "guidance" planning, decentralization of economic decision-making power from state planning and administrative agencies to productive enterprises, and the expansion of market mechanisms (for instance through price deregulation, creation of a credit system and capital markets, and the use of legally enforceable contracts between suppliers and users). In this paper, I am interested in the extent to which these reforms find expression in the current theory and reality of the Cuban planning system and their potential relevance to Cuba's evolving development strategy for the rest of the 1980s.

The empirical basis of the paper is limited[1] and any conclusions I draw are tentative and must be viewed with caution. However, given the lamentably weak state of our knowledge of Cuban planning at present, it is hoped that this paper may go some way towards plugging a yawning hole. It may also serve to stimulate more systematic work in the future.

2. SOME BACKGROUND

Estimates of Cuban developmental performance over the past 25 years are hard to obtain with precision and are the subject of much debate among Western specialists (for example, Brundenius and Zimbalist, 1985). The overall picture, however, seems to be one of very creditable performance on a number of vital aspects (economic, social and political) in an international context which has been strongly unfavorable (the enormous constraints imposed by the United States blockade), but with several basic economic problems remaining unresolved (for overviews of Cuban development performance, see Brundenius, 1984; Carciofi, 1983; Zimbalist and Eckstein, 1986; Horowitz, 1984; and Mesa-Lago, 1981). In terms of economic growth, the 1960s were disappointing but performance in the 1970s and 1980s has been respectable, albeit somewhat uneven. The industrial sector has gradually expanded and both agriculture and industry have become more diversified. In social terms — i.e. maintaining a relatively egalitarian income distribution and relatively full employment and providing a high standard of basic welfare needs, notably in education and health — Cuba has performed impressively, a fact attested by observers across a wide ideological spectrum (although housing received insufficient attention until recently).

On the negative side, however, certain basic economic problems remain, many of them maintained or exacerbated by the US blockade: continued dependence on sugar to generate foreign exchange and thus underpin strategic development possibilities generally; continued political and economic dependence on the Soviet Union and an irrationally high concentration of economic ties with Eastern Europe through CMEA; a considerable foreign debt in hard currency which has had to be rescheduled and must be paid for by improved export performance in the context of chronic balance of payments deficits in convertible currencies; slackening growth rates in the mid-1980s (the official figure was 4.8% in 1985 with 3 to 3.5% projected for 1986); and worrying levels of efficiency in the context of a centralized planned economy. These problems are interconnected and Cuba's room for maneuver is severely limited by US pressure. For example, Cuban economists complain that the expansion and diversification of foreign trade is made more difficult by a US embargo which reduces access to natural Latin American export markets, raises interest rates on commercial credit and the costs of marine transportation (ship rents, insurance, etc.) because of the blacklist for ships which carry Cuban cargo. The embargo also prevents access to the US market itself, even indirectly (for example, products made with Cuban nickel exported to a third country such as Italy are not admitted to the US).

However, Cubans realize that improvements in the efficiency and effectiveness of the economic planning system are one important means to tackle these problems. Cuban planning has a checkered history. During the 1960s, an unsuccessful attempt to set up a centralized system along Czechoslovakian lines was followed first by a prolonged debate (1963–66) and then by a Guevarist system in which the formal five-year plan framework was discarded. "Mini-plans" based on priority sectors or regions became dominant and were not coordinated at the national level; relations between state agencies and enterprises became highly centralized and the latter's financial autonomy virtually disappeared; the function of national economic planning was taken over by the political leadership with the role of JUCEPLAN (Junta Central de Planificación or Central Planning Board) being severely reduced, and the use of "economic mechanisms" in macroeconomic management largely gave way to political mobilization. During 1966 to 1970, Cuba could not be called a "centrally-planned" economy.

Economic performance during this "heroic" period was poor, culminating in the failure to attain the 10-million-ton sugar harvest in 1970; a reevaluation of economic strategy and planning methods began in that year. Over the 1970s there were moves in two basic directions. First, steps were taken to establish a system of "true planning," in the words of Cuban economist José Luis Rodríguez (interviewed in Havana in April 1985), i.e. a system of centrally coordinated economic administration based on "rational principles of economic calculation." This took the form of the introduction in 1976 of a new "System of Economic Management and Planning" (Sistema de Dirección y Planificación de la Economía or SDPE) — this was akin to the post-1965 marginally reformed system of Soviet central planning. Second, measures were taken to introduce a larger element of local and popular participation in processes of economic planning and management, through the introduction of a local government system known as *Podér Populár* or Popular Power and attempts to increase the role of workers and their organizations in the management of enterprises. These innovations were not brought in at one fell swoop but have been gradually introduced during Cuba's first two formal Five-Year Plans (1976–80, 1981–85) with uneven results. In this paper I shall attempt to draw some conclusions about the effects of these changes in the planning system as of the mid-1980s.

3. PLANNING IN THE MID-1980s

(a) *The national planning process*

From the vantage point of mid-1985, my central impression of the Cuban planning system is that, despite an increase in the level of central consultation with the regions, functional organizations and productive units, and very tentative steps towards decentralization of economic decision-making power and use of "economic mechanisms" over the past decade, it remains highly centralized, administrative and directive. As such, it is closer to the Soviet than the Hungarian system and contrasts sharply with the other Third World socialist economy with which I am familiar, China, where both rhetoric and reality of economic reform have been taken much further.

Let us begin with a brief description of the formal structure and functioning of the planning system. The central coordinating agency is JUCEPLAN which is charged with drawing up longer-term global plans and presenting them to the political leadership. This is a relatively small organization with only 200 professional staff (mainly economists).[2] It draws up annual, five-year and long-term perspective plans and is organized into departments or "directorates" responsible for three levels of analysis: (i) the national economy, which involves basic decisions about economic strategy, investment priorities and the balance between investment and consumption, and the calculation of overall material balances and macroeconomic projections; (ii) broad economic sectors, such as industry, foreign trade, construction, agriculture and stockbreeding, transport, communications, and commerce; (iii) branches (*ramas*), within industry for example, textiles, sugar, food etc. (each in turn divided into sub-branches, *subramas*). This is a structure familiar to students of Soviet-type central planning systems and reflects the influence of Czechoslovak advice in the early 1960s and Soviet advice in the 1970s.

JUCEPLAN organizes an *annual planning process* in the year prior to the target year involving four stages: First, JUCEPLAN, in coordination with the sectoral ministries (who submit information and requests for resources for their constituent enterprises), does a preliminary draft plan with a set of provisional "control figures" or "preliminary directive plans" at the macro level based on a material balance matrix. Second, this draft goes "down" to the ministries, to the enterprises or "combines" (*combinados*) under each ministry and to workshops within each enterprise for a generalized process of discussion about the acceptability and feasibility of plan targets and the resources needed to achieve them. Each individual component discusses only its own part of the plan: for instance, said a JUCEPLAN official, we do not discuss balance of payments deficits with factory workers. Organizations such as ANAP (the national organization of small farmers), and the trade unions are included in this consultative stage (at national and local levels), as are regional and local organs of *Poder Popular* (the system of popular representatives). Each regional and functional agency will have its own planning office which draws up plans each within its own sector and attempts to mesh these with overall JUCEPLAN priorities.

Third, comments and counterproposals are then forwarded to JUCEPLAN through the relevant central organs (notably the ministries). For example, the manager of a building materials factory in Cienfuegos said that JUCEPLAN, through their superior ministry, asked them to raise output of prefabricated building sets from 760 in 1984 to 1,000 in 1985. The factory organized a discussion of the target among its 260 person workforce and sent up a series of requests for the resources thought necessary to achieve it (this involved more workers since the factory had a chronic labor shortage and could not run a double-shift in consequence; indeed the plant was only working at about half its hypothetical capacity). JUCEPLAN makes any necessary adjustments and presents a definitive version of the annual plan to the Party leadership and, for formal ratification, to the National Assembly of the *Poder Popular*. Fourth, the plan then becomes law and is binding on all subordinate units; the "democratic side is now finished" and it reaches basic-level enterprises in the form of precise targets elaborated (and often enlarged) by their ministerial superiors.

This process, like the planning structure itself, is said to be an economic expression of the Leninist principle of "democratic centralism" (for a full Cuban description, see JUCEPLAN, 1981). Our key concern, of course, should not be with a formal description of the planning structure and process but with their real dynamics. How does this system of economic coordination and control actually work and how does it compare with socialist planning systems elsewhere? A systematic answer to these questions would require deeper study, but there is enough evidence to allow some initial judgments.

Clearly JUCEPLAN is by no means the commanding presence it appears. For most of its history in fact (up to the mid-1970s) it was either ineffectual or bypassed (Mesa-Lago, 1981, Chap. 2). It occupies an awkward position sandwiched between the demands of the Party leadership on one side and the pressures of economic and social agencies on the other. Among the latter, the central ministries in Havana clearly wield a good deal of power to get their spending plans approved. This is a familiar feature of socialist planning systems elsewhere (for example, compare the process of "centralized pluralism" described by Nove (1977, Chap. 3) in the Soviet Union). A dramatic expression of its difficult position occurred in late 1984 when the draft plan for 1985 came under heavy criticism from the Commandante himself, Fidel Castro, who in effect told the planners to go back and change the way they operate. The occasion was a final high-level meeting in late November to approve the 1985 plan, attended by members of the Party's Politburo and Secretariat, vice-presidents of the government's Executive Committee, economic and other ministers; provincial Party secretaries and *Poder Popular* presidents; and heads of mass organizations such as the trade unions. If one is seeking an "economic decision-making elite" in Cuba, this group is probably it, with the Party representatives playing the dominant role within it. This meeting severely questioned the draft 1985 plan and set up a high-level working group to work "on an emergency basis in permanent session" to revise the plan (Castro, 1985, p. 34). Part of this revision involved an attempt to change the process of plan-formation by tackling its "sectional spirit." Fidel Castro described the normal pattern as follows (1985, p. 35):

> All those years, ever since we made our first efforts in planning and development, a sectional spirit prevailed in all the state agencies, in every ministry and, in the end, in practice the plan was not necessarily rational or optimal. Rather, it reflected the sum total of each sector defending its interests before the planning agencies and vying for the available resources . . . (Everybody) demanded resources and each sector claimed that its needs were the most essential, the most decisive and the most important.

Retrospectively, argued Castro, what should have been a rational process of economic coordination was in fact a war (1985, p. 35):

> It really was like that. We realised this when we reviewed the history of each plan every year. What we saw was a war being waged by each agency, a struggle, a battle for what limited resources we had.

Castro demanded a new spirit of commitment to the national economy as a whole. This view was echoed by Felino Quesada of JUCEPLAN in response to a question about projected changes in plan procedure; he said that they were aiming for more participation and a more "collective" spirit among the different agencies.

(b) *Changes in the planning process: Collectivism and participation*

Two points should be raised here: first, about the difficulty of achieving this "collective spirit" and, second, about the potential contradiction between this and the priority of greater participation in plan making.

It is difficult to envisage how the new spirit of economic collectivism can be instilled. Certainly, no concrete method has been offered so far, except promises that each claim on resources would be subjected to greater scrutiny and the ensemble subjected to more rigorous discipline from above (enforced by the new central working group). This includes greater control not merely over ministerial budgets but also over investment programs organized by local organs of *Poder Popular* (Castro 1985, pp. 40–41). Clearly the Party (with Fidel at the apex) plays a crucial coordinating role throughout the planning system, intervening to enforce overall priorities and mediate conflicts (for example, between a ministry and one of its enterprises). Party organs at the local level, moreover, clearly play an important role in sorting out some of the characteristic problems of a centralized directive planning system. For example, local Party leaders might arrange for alternative sources when planned supplies fail to arrive at a factory within their bailiwick, or enforce technical cooperation between administratively unconnected plants. For instance, production in the building materials factory in Cienfuegos was restricted by the lack of metal moulds and they were trying to make substitutes themselves. They requested help from local factories but some "pretend they haven't heard"; but the Party group in the local construction industry intervened to secure their cooperation.

There is also the familiar Cuban political approach to the problem of economic sectionalism, fostering a new "economic mentality" to replace the old. Castro describes the latter as follows (1985, p. 54):

> (We always assumed that) everything was available always without asking ourselves how it all came about. (This) gave shape to a wasteful mentality, a mentality of little thrift, a consumption rather than export mentality, an import rather than export mentality.

However if the record of the 1960s showed one thing in Cuba it was the limitations of a policy of transforming underlying structural problems by means of moral suasion and political mobilization, even though the "heroic" approach was justifiable in the context of the times (see Zimbalist and Eckstein, 1986). Since the process of "centralized pluralism" is rooted in the basic institutions of the directive planning system, it is hard to see how a new collective spirit can be achieved, except at the margin or in the very long term. The problems of "sectionalism" lie rooted in the institutional logic of this type of planning system. As such, they are structural problems which need structural solutions; this is a major part of the case put by economic reformers in other socialist countries who argue that this economically irrational process of bureaucratic politics which passes for "central planning" can be alleviated by transferring power from state institutions to enterprises. *Prima facie* this argument seems applicable to the Cuban situation.

We now turn to the question of *participation* in the planning process. It appears that the Cuban government has made a serious attempt to institutionalize processes of consultation and discussion at all levels (Carciofi, 1983; Cockburn, 1979). Though the electoral principle underlying *Poder Popular* institutions is limited (direct election only operates at the basic "precinct" level), and their economic (as opposed to social) role appears very circumscribed, it seems likely that *Poder Popular* does give some expression to regional and local interests within the planning process, as well as providing outlets for citizen complaints on mundane matters. At the factory level, moveover, there seem to be serious attempts to bring the workforce into discussions about the feasibility of the initial control targets but it is unclear precisely how much difference this makes to ultimate decisions. My impression is that in comparison with other state socialist economies there may be a relatively high degree of consultation and discussion at all levels of the Cuban planning system. Thus Zimbalist is right when he argues (1985, p. 215) that "planning in Cuba has evolved in a more decentralized and participatory fashion than in the Soviet Union" even though the structure of their planning systems are similar. But this kind of participation does not entail a significant diffusion of power nor should it be confused with democratic control. It still takes place within a hierarchical framework of Leninist political economy and the impact of levels below the center appears to be marginal. At the regional level, planning capacities do not appear to be well developed as yet and this may weaken the capacity of regional governments to press their case in Havana; the independent economic power of local governments (*Poder Popular*) also seems very circumscribed. Indeed, officials of the Cienfuegos *Poder Popular* admitted that "economic power is centralized at the ministry level" and, though their regional needs were "taken into consideration" in the national plan, the main economic function of the local *Poder Popular* was to "fulfil the norms" of central agencies (just as their central political function was to "fulfil the Party's norms"). (For further discussion of the restricted economic role of *Poder Popular* organizations, see Slater 1986, pp. 14–18.) As such, local governments may be weaker than their counterparts in the larger state socialist countries, China and the Soviet Union.

The government seems committed to extending the principle of popular participation, direct or representative, in the planning process. Quite apart from being an aim valuable in itself, the extension of socialist democracy, it has clear potential benefits for the planning process in terms of increasing flows of information and increasing people's commitment to plan targets by enlisting their prior involvement and consent. At the same time, however, "real" participation is not merely a process of rational discussion of economic objectives, but a political process which involves the airing of opposing views and competing interests. Its economic results may be very ambiguous.

The planning process is in fact a complex matrix of bargaining between institutions, levels and groups, each constituency no doubt structuring the "facts" to support its case (whether this be workers' desire for a lower work-norm, a factory's desire for lower plan targets or more resources, or a ministry's application for a budget increment). This process involves bureaucratic agencies (notably the branch ministries), regional/local governments, corporate organizations, mass organizations and productive enterprises. If we focus on the latter, in agriculture the management of state farms and the executive committees of cooperatives bargain with state agencies each year over credit, quotas and prices; for the private sector, the National Association of Small Farmers (ANAP) is important as a bargaining agency. Turning to industry, the manager of a Cienfuegos building materials plant gave some of the flavor of the bargaining game between plant and ministry. The factory sent a team to negotiate targets and resources with the ministry and reported on success to their workers' council: "We tell them what we won and what we lost, whether we were lucky or unlucky" (he added that the workers were sometimes unhappy at the outcome and criticized management for not putting up a better performance). This sense of a common adversary, the ministry, tends to create certain common interests between plant managers and workers, though the director himself is caught in the familiar "foreman's dilemma" between the demands of his administrative superiors and his own workforce.[3]

Expanded participation in the planning process is thus problematic on several grounds. To the extent that it leads to a further politicization of the economy, the economic results may not come up to expectations. Moreover, at a time when the Party leadership is trying to impose more stringent discipline on the planning process (including expenditure cuts which may damage established interests), the extension of real as opposed to ritual participation is clearly problematic. Participation is also problematic if it is being fostered as an *alternative* to more thoroughgoing economic reforms based on an extension of market relations (I discuss the case for the latter below). It is hard to see how the complex, competitive political process which real participation would involve can solve the efficiency problems of a traditional central planning system without a concomitant expansion of market relations. The hypothetical claim of greater participation leading to greater involvement and commitment leading to higher productivity may have weak links if most work remains a disutility and the enterprise is not subject to external pressures to perform, both positive and negative. If "vertical" bureaucratic pressures are to be removed, therefore, "horizontal" market pressures become more important. Moreover, the expansion of market relations entails a radical decentralization of economic decision-making power to the enterprise; without this, worker participation within enterprises lacks substance. Both political and economic goals would seem to require that market relations and worker participation be extended simultaneously.

(c) *Changes in the planning system: Decentralization and economic reform*

To sum up our discussion so far, the Cuban planning system is highly centralized in two senses: the central organs dominate over regional and local levels and the process of participation takes place within a hierarchical framework. It is also highly centralized in terms of a third dimension, the relationship between the state agencies of economic planning and administration on the one hand, and basic-level productive units on the other. In spite of limited moves to increase the decision-making powers of enterprises over the past few years, their degree of independence still seems very limited. Correspondingly, relations between state agencies are based on administrative directives with only marginal use of "economic mechanisms" such as taxation, credit, price policy, etc. From this point of view, Cuba resembles the (still basically directive and only marginally decentralized) Soviet system rather than the Hungarian system (which uses indirect planning methods and where decentralization has been carried much further).

Though enterprises have their own accounting systems and are supposed to maximize net revenue, the bulk of their activity is regulated by mandatory plan targets. These are based on JUCEPLAN's general directives but which are specified, added to and transmitted by the relevant superior ministries. Essentially enterprises are told what to produce, at what prices, for what outlets, what raw materials to use, and how much to pay their workers. Investment funds are centrally controlled and most of enterprise profit is claimed by the state.

It is clear that the Cuban economy is plagued by many of the characteristic problems of centralized directive planning systems which have been identified by reformers elsewhere as evidence of the need for change: systemic "scarcity" (in the sense analyzed by Kornai, 1985), widespread waste, absenteeism, "unemployment on the job," weak incentive systems, inefficient use of investment, "storming" to meet targets by the end of the month or some politically significant date (such as 26 July) and so on.

Cuban planners and economists are aware of these problems and of their relationship to the relatively high degree of centralization in the planning system. However, they tend to defend the latter on several grounds: First, the relatively low level of development of Cuba requires central control to mobilize and allocate resources, in contrast to more advanced economies in Eastern Europe where parametric planning methods and market-type reforms might be more appropriate. Since underdevelopment in a small country entails structural imbalance and international dependence, strong central controls are necessary for restructuring the domestic economy and changing its relationship to the international economy. They are also necessary to correct inequalities between regions. Second, international political-economic hostility also requires centralization, notably the US blockade and fears of political subversion and economic sabotage (for example, Cubans suspect that the agriculture blights of 1980–82 which damaged sugar, tobacco and livestock production, were not accidental). Heavy military demands on the economy strengthen the rationale and the reality of centralization.

One might add further reasons. First, the earlier commitment of the Cuban revolution towards radical socioeconomic redistribution provided a strong case for control from Havana and created a large machine to administer it. The existence of this large bureaucracy may itself pose an obstacle to economic decentralization. Second, given the relatively small size of the economy, the problems of directive planning systems may be less than in larger countries, at least in the early stages of growth.[4] Third, the political logic of revolutionary mobilization, which is still important in Cuba, creates an ideological atmosphere which reinforces control by the central Party leadership and continues to regard markets as "bourgeois" Trojan Horses (for instance, see Castro's speech condemning free peasant markets as agents of harmful "commercialism and speculation": *Granma Weekly* (1 June 1986). Fourth, heavy economic and political reliance on the Soviet Union may extend to the use of Soviet planning institutions. It is worth recalling, moveover, that "central planning" in any true sense was not set in train in Cuba until the mid-1970s and has only recently been consolidated; reforms of the Hungarian type might thus appear premature.

The tentative reforms in economic management introduced are relatively conservative in that they wish to extend commercial mechanisms without sacrificing central controls. For example, there have been some steps towards making use of (repayable, interest bearing) credit instead of budgetary grants to finance enterprises. In the past, this has been limited to working capital but, starting in 1985, the principle is being extended to investment finance. Where greater power has been granted to enterprise managers, it appears very limited and qualified. Successful managers do have the right to retain part of their annual profit to pay individual and collective bonuses, but this is set at only a small proportion of total profit. In INPUD in Santa Clara (Instituto Nacional de Producción de Uténsiles Domésticos), 10% of net profit went for bonuses and 13% for special bonuses for outstanding workers; of the remaining 77%, most went to the state but part was retained for welfare expenditure and reinvestment. However, INPUD could not launch its own investment program; this had to be approved by the Ministry of Metallurgy and appropriate funds allocated from the national budget. In general, however, according to Felino Quesada of JUCEPLAN, retained funds are relatively small; most net revenue goes to the state. Even where enterprises have obtained control over significant financial resources for distribution and investment, they have been unable fully to realize them given the lack of reform in other aspects of the economy and a general context of scarcity, real or induced.

Further economic reforms are probably in prospect in the near future; these were to be discussed and ratified at the Party's Third Congress in early 1986 (as of writing this article, I have not seen the documents from this Congress). Cuban economists with whom I talked certainly favored further reforms in a market/decentralization direction and hoped for significant innovations at the Third Congress; some, for example, hope for a systematic reform of both producer and consumer price systems before the end of the decade. The rate and extent of economic reforms is clearly the focus of vigorous debate. One might point to certain features of, and trends in, the Cuban economy which would strengthen the case for more basic reforms, perhaps along Hungarian or Chinese lines. First, the economy is still dominated by agriculture (with about 80% of land in state hands). Comparative experience suggests that traditional central planning is particularly ineffective in regulating this sector given its climatic uncertainty and the particularities of agricultural work processes (Fitzgerald, 1985). Cuban economists readily admit that Cuba's record in agricultural planning leaves much room for improvement, particularly in the non-sugar sector, and repeat the joke that the official most frequently replaced in Cuba, as in the Soviet Union, is the Minister of Agriculture. Second, given the switch in developmental priorities in the 1970s away from redistribution towards growth and productivity, the case for a more decisive move towards parametric planning and economic decentralization would be stronger. Third, to the extent that economic performance and structural diversification were reasonably successful in the 1970s, the economy has become more complex and thus more resistant to direct central control. To this extent, the argument that economic decentralization is less necessary because of the relatively small size of the Cuban economy is weakened (notwithstanding the apparent illusions of some Cuban planners about the potential power of computers in reducing informational problems). Fourth, to the extent that repayment of Cuba's hard currency debt (US $3 billion) and the future dynamism of the Cuban economy depend on the country's ability to compete in international capitalist markets, competitive pressures abroad may force the pace of domestic economic reform. Successful export performance, as the experience of places like Hong Kong has shown, requires an ability to monitor changes in international demand and to create new products or switch production lines accordingly. Though Cuban planners are trying to encourage this flexibility, it may well be difficult within the present framework.

However, the exigencies of international political isolation and military threat may well outweigh the purely economic rationale for more thoroughgoing reforms. The continued ideological hostility of the political leadership towards markets still poses a potent constraint on economic liberalization; witness the decision to abolish the (limited) free peasant markets in mid-1986 (*Granma Weekly*, 1 June 1986). While these countervailing political pressures may blunt the impetus for change, it is to be hoped that in the wake of its Third Congress in February 1986 the Party will continue to move planning reform in the direction it has been taking, albeit slowly and unevenly, since the mid-1970s.

NOTES

1. The analysis is based on material collected during a two-week visit to Cuba as a member of a specialized group, organized by the Architectural Association's School of Planning. The group, which contained specialists on national and regional planning, followed a schedule of visits and interviews in Cuba which were chosen to throw light on the various aspects of economic and physical planning.

The author would like to thank the other members of the group for their stimulating company and their commitment to systematic inquiry. As a novice in Cuban affairs, I learnt a great deal from fellow group-members and would like to thank David Wield (Open University) for his comments on an earlier draft of this paper. The views expressed here are the author's own and should not be identified with the AA School of Planning.

The author would also like to thank Charles Legge of the Britain–Cuba Resources Centre and Jean Stubbs for their invaluable advice and help in arranging a very fruitful itinerary.

2. JUCEPLAN is also responsible for four attached institutions: the Institute of Physical Planning, concerned with spatial dimensions of the economy and regional policy; the Institute of Economic Research which mainly handles long-term strategic planning and forecasting; an Economic Directorate which trains management cadre for the rest of the economy and provides short-term refresher courses for senior ministerial officials; and, somewhat incongruously, the National Office of Industrial Design which may be hived off in future.

3. This Cienfuegos plant was a small-to-medium size concern (260 workers) and we should consider carefully the implications of scale for the role of an enterprise in processes of planning and participation. We can divide industrial enterprises roughly into two categories: (i) big enterprises or "combines" (composed of horizontally or vertically integrated plants) which dominate certain sectors or subsectors (there are about 200 of them); (ii) "ordinary" or medium to small enterprises, with few workers, simpler structures and smaller output capacity. One would expect, first, that big enterprises would have much greater leverage in bargaining with their administrative superiors than ordinary enterprises; second, that the process of intra-firm consultation and participation is more institutionalized and formal in big enterprises and more informal in ordinary enterprises. My superficial impressions of ordinary and big firms (e.g., INPUD in Santa Clara with 2,250 workers) tend to support these hypotheses.

4. The argument that smaller *size* makes central planning easier can be questioned on several grounds. First, the strongest economic reform currents in Eastern Europe have come from smaller countries (Czechoslovakia in the 1960s and Hungary). Cuba has about the same population as Hungary and, when compared with its neighbors in the Caribbean at least, could hardly be called a "small" economy. Second, though the informational problems posed by complexity would be reduced in a smaller economy, one would still be talking about different levels of manageability (unless one descended to the minuscule), especially in poorer societies where infrastructure and communications are not well developed. Third, the characteristic problems of central planning operate to a large degree independent of size, *within* the component subsystems of a large complex planning system as well as in the whole. In consequence, attempts at raising economic performance through regional decentralization (for example, in China and the Soviet Union) have not proven successful.

REFERENCES

Brundenius, Claes, *Revolutionary Cuba: the Challenge of Economic Growth with Equity* (London: Westview Press, 1984).

Brundenius, Claes, and A. Zimbalist, "Recent studies on Cuban economic growth: A review," *Comparative Economic Studies*, Vol. 27, No. 1 (Spring 1985), pp. 21–45.

Carciofi, Ricardo, in White *et al.* (1983), pp. 193–233.

Castro, Fidel, *This Must Be an Economic War of All the People*, Speech 28 December 1984 (Havana: Editora Politica, 1985).

Cockburn, Cynthia, "People's power," in Griffiths and Griffiths (Eds.) (1979), pp. 18–35.

Feuchtwang, Stephan, and Athar Hussain (Eds.), *The Chinese Economic Reforms* (London: Croom Helm, 1983).

Fitzgerald, E. V. K., "The problem of balance in the peripheral socialist economy: A conceptual note," in G. White and E. Croll (Eds.) (1985), pp. 5–14.

Griffiths, J., and P. Griffiths (Eds.), *Cuba: The Second Decade* (London: Writers and Readers Publishing Cooperative, 1979).

Halebsky, S., and J. M. Kirk (Eds.), *Cuba: Twenty-Five Years of Revolution 1959–1984* (New York: Praeger, 1985).

Horowitz, I. L. (Ed.), *Cuban Communism*, 5th edition (London: Transaction Books, 1984).

JUCEPLAN, *El Sistema de Direccion y Planificacion*

de la Economia en las Empresas (Havana: Editorial de Ciencias Sociales, 1981).

Kornai, Janos, "Comments on papers prepared in the World Bank about socialist countries," *CPD Discussion Paper*, No. 1985-10 (Washington, DC: March 1985).

Mesa-Lago, Carmelo, *The Economy of Socialist Cuba: A Two-Decade Appraisal* (Albuquerque: University of New Mexico Press, 1981).

Nove, Alec, *The Soviet Economic System* (London: Allen & Unwin, 1977).

Nove, Alec, *The Economics of Feasible Socialism* (London: Allen & Unwin, 1983).

Slater, David, "Socialism, democracy and the territorial imperative," Mimeo (Amsterdam: CEDLA, 1986).

White, Christine, "Recent debates in Vietnamese development policy," in G. White *et al.* (1983), pp. 234-270.

White, Gordon, and Elisabeth Croll (Eds.), "*Agriculture in socialist development*," special issue of *World Development*, Vol. 13, No. 1 (January 1985).

White, Gordon, Robin Murray, and Christine White (Eds.), *Revolutionary Socialist Development in the Third World* (Brighton: Harvester Press, 1983).

Zimbalist, Andrew, "Cuban economic planning: Organisation and performance," in Halebsky and Kirk (Eds.) (1985), pp. 213-230.

Zimbalist, Andrew, and Susan Eckstein, "Patterns of Cuban development," *World Development*, Vol. 15, No. 1 (1987), pp. 5-22.

·11·

Trade, Debt, and the Cuban Economy

RICHARD TURITS

1. INTRODUCTION

In August 1982, The Cuban National Bank announced its inability to meet debt obligations to the West, and requested a rescheduling on its principal payments due to mature between 1982 and 1985. By that time, Cuba had amassed a hard currency debt over $3 billion, equal to roughly three-and-a-half times the value of exports to the West. In 1982, interest alone amounted to some 45% of the value of exports to the West (Unctad, 1982, Anexo III, p. 15). Terms of trade, moreover, had deteriorated severely as the world market price of sugar fell to a level below the cost of production. The debt crisis in Cuba coincided with the specter of a Mexican default. And the constraints faced by Cuba seemed remarkably similar to those of the rest of Latin America. Despite Cuba's socialist polity and tight integration into CMEA, the island still appeared profoundly influenced by the vicissitudes of the world market, and still dependent on Western capital — recalling what one writer had recently described as the "capitalist constraints on socialist development in Cuba" (Eckstein, 1980).

Yet the analysis suggested in the context of the debt crisis now deserves reconsideration in light of the unexpected ease with which Cuba has managed its hard currency obligations. Cuba has met its interest payments on time, apparently without precarious cuts in social welfare spending, or substantially-reduced growth rates. Indeed, Cuba has witnessed a build-down of its debt probably unique in Latin America. Commercial debt reported to the Bank for International Settlements (BIS) decreased from $2.1 billion in 1979 to $0.9 billion in 1982.[1] Though the total debt has increased slightly since 1982, Cuba's debt in 1984 was 5% below its 1981 level, while that of the rest of Latin America had risen 30% (see Table 1 and *Latin American Weekly Report*, 10 May 1985).[2] Further, as the rest of Latin America experienced politically-destabilizing austerity programs, and negtive real growth rates, Cuba increased spending on social welfare, widened the availability of consumer goods (see *Latin American Regional Report*, 30 September 1983), and simultaneously maintained robust growth rates (see Table 2). If we imagined Cuba to be still deeply entrenched in a "world capitalist system," *à la* Immanuel Wallerstein (see Eckstein, 1980), this outcome would have been unpredictable given the near seismic deterioration in Cuban terms of trade — as oil prices soared and the sugar market dwindled — and given Cuba's position in 1979 as the Latin American country with the fifth largest debt in proportion to exports (Eckstein, 1985, Table 4).[3]

The question raised by Cuba's surprising strength is whether it reflects Cuba's increased dependence on Soviet aid, the peculiar characteristics of Cuba's socialist polity, or some combination of the two. In answering this question, I will narrate the history of the Cuban debt, discuss

---

*Thanks to Edward Ames, Miguel Centeno, Julie Feinsilver, Hernando Gómez, Laura Gotkowitz, Robert Kaufman, Michael Marrese, Barbara Stallings, Konrad Stenzel, and especially Andrew Zimbalist for helpful comments and criticisms on earlier drafts of this article. Of course, responsibility for the article is entirely my own.

Table 1. *Foreign debt in convertible currency*

Year	Amount (million pesos)	BIS-reported debt ($ million)*	BIS-reported short-term debt ($ million)	Total financial debt (million pesos)	Total short-term debt (million pesos)	Official bilateral debt (million pesos)	Multilateral aid (million pesos)
1978		1,836	1,035				
1979	3,267	2,080	1,176	1,953	1,269	1,280	
1980	3,227	1,834	1,016	1,837	1,238	1,354	8
1981	3,170	1,572	833	1,826	1,282	1,294	15
1982	2,669	869	384	1,327	860	1,276	18
1983	2,790	1,031	580	1,335	789	1,333	25
1984†	3,033	1,027	586	1,264	699	1,526	20

Sources: BIS (various issues); BNC (February 1985).
*Includes loans from commercial banks in group-of-ten countries, Switzerland, Austria, Denmark and Ireland, and certain of their affiliates.
†Preliminary figures except for BIS columns.

Cuban economic development, and analyze the nature of "socialist dependency." First, I will examine the origins of Cuba's debt problems, and the policies which were followed to restore external balance. Was the 1982 liquidity crisis the product of internal deficiencies, external shocks, or other structural problems? This question will be addressed by analyzing Cuba's current and capital accounts from 1974 to 1982, and concomitant economic policy and conditions.

Interestingly, the build-down of Cuban debt began in 1980 before any immediate threat of a liquidity crisis appeared, a move suggesting a somewhat voluntary policy-making decision (Unctad, 1982, p. 16), and perhaps reflecting an overall CMEA strategy to build-down Western debt where possible. While total Latin American debt reported to the Bank for International Settlements more than doubled between 1978 and 1982, Eastern European debt increased only marginally from $48 to $53 billion (BIS). Cuba's debt was thus more of a piece with other CMEA nations than with the rest of Latin America. In fact, the actual liquidity crisis in Cuba in 1982 was provoked three years after external balance was restored with market economies, and after Cuba had already begun to reduce its hard currency debt (see Table 3).[4] The crisis was thus largely the product of new international financial policies, i.e., the general contraction in private bank lending to the Third World, that, in Cuba's case, occasioned especially large outflows from the capital account.

Next, I treat perhaps the most striking aspect of the Cuban debt, how Cuba has been able to pay its debt without traditional austerity measures or economic recession. The deterioration in Cuba's terms of trade, and the island's increased debt with the West have been "capitalist constraints" in the sense that they have pushed Cuba away from trade with the West and into further commerce with CMEA. Analogous to Cuba's nearly inevitable response to the US embargo, Cuba has substituted capitalist and US constraints with increased CMEA integration. As will be discussed, that integration has been on increasingly favorable terms relative to world market prices. The extent of implicit subsidies[5] is controversial and the logic of Soviet policy is far from clear. Nonetheless, I will speculate on the nature of Soviet dependence, socialist aid, and the latter's tied and convoluted form — the last exemplified by the fact that Cuban re-exports of Soviet oil have become the island's foremost method of obtaining foreign exchange. I conclude that Cuba's socialist polity and Soviet aid are inextricably linked as necessary conditions of Cuba's strategy for economic development and debt without austerity.

2. ORIGINS OF THE WESTERN DEBT: CURRENT ACCOUNT DEFICITS

In the 1960s, few Western loans had been available to Cuba, which belongs neither to the World Bank nor the Inter-American Development Bank. In 1969, Cuba's debt was a mere $291 million (Banco Nacional de Cuba (BNC), August 1982a, p. 13). In the early 1970s, the international financial outlook changed radically

Table 2. *Cuban growth rate* (in constant prices)

Year	GDP*	GDP per capita	GSP†
1959	5.0	2.9	
1960	1.4	−0.4	
1961	4.0	2.5	
1962	−1.4	−3.0	
1963	1.7	−0.6	
1964	9.5	6.7	
1965	0.9	−1.6	
1966	−1.9	−4.1	
1967	2.2	0.2	
1968	6.3	4.4	
1969	−2.7	−4.3	
1970	−0.5	−1.9	
1971	−0.7	−2.3	
1972	8.4	6.4	
1973	12.8	10.5	
1974	8.7	6.9	
1975	10.5	8.7	
1976	8.3	6.7	
1977	8.9	7.5	
1978	6.1	5.0	
1979	2.3	1.6	
1980	2.3	2.7	3.3
1981	9.3	8.5	13.9
1982		1.3	2.5
1983			5.2
1984			7.4
1985			4.8

Sources: Brundenius (1984, pp. 40, 121); EIU (1986, No. 1, p. 2).
*GDP figures are estimated by Brundenius by adding Total Material Production — which, unlike GSP, does not doublecount intermediate inputs — and health and education, which serve as a proxy for all non-material services (Brundenius, 1984, pp. 19–41).
†Global Social Product. GSP doublecounts intermediate inputs and excludes non-material services (primarily education and health). This is the official national income statistic.

for Cuba as most Western nations began to renew relations with Cuba. For example, Argentina signaled rapprochement with a $1.2 billion export credit to Cuba when Perón returned to power in 1973. And in 1975, the Organization of American States lifted its sanctions against diplomatic relations with Cuba.

In the wake of the spectacular failure of the 10-million-ton sugar harvest in 1970, and disillusionment over the potential for sugar to usher in the foreign exchange necessary for industrial takeoff, Cuban leaders eagerly utilized credit and capital when it suddenly became available. Despite its anomalous position as the only socialist country in the Western hemisphere, Cuba was included in the burgeoning borrower's market that resulted from the huge OPEC surpluses which needed to be circulated. The inaccessibility of the far more generous, though politically more constrained, multilateral lending,[6] and the absence of foreign equity investment — so important to pre-revolutionary Cuba — meant an overdetermined recourse to commercial bank lending, as, in fact, was typical for CMEA countries (Theriot, 1979, pp. 179–180).[7] While the Soviets had little hard currency to offer, the Eurmomarkets relayed increasing amounts of foreign exchange and credits. Charges to Cuba of 1.75% above Libor were the highest among CMEA nations — countries which are seemingly secure investments, given the planned nature of their economies, the domestic political control, and the bankers' expectations of a Soviet guarantee — but about average for less developed countries (Theriot, 1979, p. 184).

The increase in credit and capital coincided with the most dramatic climb in raw sugar prices to an historic apogee of 64 cents per pound during 1974, which in turn improved Cuba's credit rating. Given the availability of foreign exchange, and the lifting of embargos, Cuba rapidly took advantage of Western trade — its lower shipping costs, more competitive products, and resultant decrease in dependence on the Soviet Union. Western imports nearly quadrupled in the years 1973–75. And trade with the West reached over 40% of total trade in 1975, compared with 20% in 1967, and less than 14% in 1983. Despite 15 years of attempted Soviet substitution of US technology and spare parts, Western technology continued to play a small but essential role — the enlarging of which presumably would contribute to growth, and eventually more competitive products. Castro periodically criticized Soviet goods for their poor quality, and bemoaned the fact that certain needed items remained unavailable in the socialist world. During the 1960s, for instance, it seems that hard currency constraints had led to the importation of Soviet-developed combine harvesters, rather than the more sophisticated and reliable Australian model. The Soviets had no experience with sugar cane (they grow beets) and their KCT-1 proved to be an economic catastrophe. A recent analysis by a Swedish economist concluded, "even if ithe machines were given to Cuba free — which is not likely — it was probably a very costly enterprise," delaying successful mechanization from five to eight years (Edquist, 1985, pp. 126–127). Fidel Castro described the Soviet harvester as the "great destroyer — where it had passed nothing will grow

Table 3. *Balance of trade** (at current prices, million pesos)

Year	Market economy countries Exports	Imports	Balance	Socialist countries Exports†	Imports‡	Balance	Total Balance
1960	486.4	518.7	−20.0	149.6	119.3	30.3	−20.0
1961	166.5	191.8	−25.4	458.2	446.8	11.4	−14.0
1962	93.7	130.4	−36.7	427.0	628.9	−201.9	−23
1963	177.2	163.7	13.5	366.6	703.6	−337.0	−323.5
1964	291.3	331.6	−40.3	422.5	687.2	−264.7	−305.0
1965	149.4	207.8	−58.4	536.1	658.4	−122.3	−180.7
1966	110.3	186.6	−76.3	482.2	738.9	−256.7	−333.0
1967	134.8	209.2	−74.4	582.2	791.8	−209.6	−284.0
1970	272	406	−134	778	905	−127	−261
1971	296.6	417.6	−121.0	564.6	469.9	−405.3	−526.3
1972	349.1	274.9	74.2	421.8	914.9	−493.1	−418.9
1973	408.5	427.4	−18.9	744.5	1,035.2	−290.7	−309.6
1974	952.9	873.7	79.2	1,283.6	1,352.3	−68.7	−10.6
1975	949.4	1,508.5	−559.1	2,002.8	1,604.6	398.2	−160.9
1976	602.3	1,315.5	−713.2	2,090.0	1,864.2	225.8	−487.4
1977	475.1	1,121.0	−645.9	2,443.3	2,340.6	102.7	−543.2
1978	524.2	724.8	−200.6	2,915.9	2,849.0	66.9	−133.7
1979	615.4	634.5	−19.1	2,883.9	3,053.0	−169.1	−188.3
1980	1,180.5	1,005.7	174.8	2,786.2	3,539.6	−753.4	−578.6
1981	1,044.6	999.8	44.8	3,179.2	4,114.2	−935.0	−890.2
1982	760.1	628.3	131.8	4,179.5	4,908.8	−729.3	−597.5
1983	768.8	814.6	−45.8	4,753.9	5,403.1	−649.2	−695.0
1984	569.3	1,148.9	−579.6	4,892.8	6,058.3	−1,165.5	−1,745.1

Sources: CEE (1982); BNC (December 1982b; December 1983; February 1985); Roberts and Hamour (1970).
*Export shipments are valued on FOB terms, and imports on CIF terms.
†Includes exports paid for in convertible currency.
‡Includes imports from market-economy countries paid for in transferable rubles.

for a long time to come" (Radell, 1983, p. 374).

In an interview during the mid-1970s, Deputy Prime Minister of Foreign Affairs Carlos Rafael Rodríguez explained the reasons for the trend toward Western trade:

> Cuba's importations of raw material and industrial equipment from nonsocialist countries are increasing today as our investments expand. This opening of trade with the West is also occurring in Eastern Europe and for the same reason: a whole range of technology is still not available in the socialist camp. (*Le Monde*, 16 January 1975, p. 4)[8]

Though a clear breakdown of trade differentiating between socialist and capitalist countries is not available in Cuban statistics, the development-oriented direction of these imports is discernible. Western imports consist almost entirely of capital and intermediate goods, with the latter increasing over time as hard currency diminished, price shifts occurred, and Cuba became more self-sufficient in the production of capital goods. Between one-third and one-half of Western imports comprised machinery and transport material between 1978 and 1981. Chemicals and their byproducts were even more overrepresented relative to total trade, and constituted some 10% of Western imports during the same period (Unctad, 1982, Anexo, pp. 9–10). In 1983 and 1984, capital and intermediate goods reached 93% of hard currency imports, and intermediate goods alone represented roughly 70% (BNC, 1985, p. 37).

Cuba's capacity to import Western goods with domestic funds, however, depended on ephemeral increases in the prices of sugar. After climbing from 3.7 cents per pound in 1970 to 64 cents at one point in 1974, the price of sugar plummeted to eight cents by 1977 (see Table 4). As the price fell, Cuba's recourse to Western goods fell accordingly. While the nominal value of Western imports had increased from 0.4 to 1.5 billion pesos[9] in the years 1973–75, the latter was more than halved between 1975 and 1979. Anxious to utilize foreign capital when available, however, Havana registered one of its largest Western

Table 4. *Sugar: prices, volume and subsidy*

Year	Free market price (cents per pound)	US import unit value (cents per pound)	Soviet import unit value* (cents per pound)	Soviet imports (thousand tons)	Estimated Soviet sugar subsidy over US import price ($ million)	Total Cuban exports (thousand tons)
1970	3.7	7.0	5.9	3,105	−75	6,906
1971	4.5	7.2	5.9	1,581	−45	5,511
1972	7.3	7.9	6.4	1,097	−36	4,140
1973	9.5	8.8	11.7	1,661	106	4,797
1974	29.7	19.5	19.4	1,975	4	5,491
1975	20.4	24.5	26.5	3,187	141	5,744
1976	11.6	12.6	27.6	3,036	1,004	5,764
1977	8.1	9.2	28.0	3,790	1,571	6,238
1978	7.8	8.6	36.2	3,936	2,395	7,197
1979	9.7	10.1	36.8	3,842	2,262	7,194
1980	28.7	22.3	47.5	2,726	1,515	6,170
1981	16.9	21.2	35.1	3,204	982	7,055
1982	8.5	15.3	33.0	4,426	1,727	7,727

Sources: CEE (1973; 1975; 1979; 1982); Unctad (1982, Anexo, III, p. 17); US Department of Commerce *Statistical Abstract of the United States* (various issues).
*The official peso–dollar exchange rate is used to convert unit values.

deficits already in 1975 — when its Western export earnings were still at a peak; and when the island's high credit rating permitted prodigious borrowing. And as the needed retrenchments in imports lagged behind the sharp decline in exports to the West, Cuba accumulated a redoubtable Western debt. The four years 1975–78 culminated with a trade deficit with market economy countries of 2.1 billion pesos — compensated for by the rapid expansion of its Western debt. Emblematic of the crisis, in 1977 an estimated 100 million dollars worth of export goods lay on Japanese docks because Havana could no longer pay for them (Eckstein, 1980, p. 267).

The rapid pace of Western import cuts, however, brought a restoration of external balance by 1979 (a deficit of merely 19.1 million pesos in merchandise trade). Despite the deficits and fallen sugar prices, lines of credit in 1977 and 1978 remained open as the short-term loans which constituted a precarious two-thirds of all private bank loans were rolled over. Funds remained available at 1.75% over Libor at a time when margins in general were narrowing (Economist Intelligence Unit, 1978, No. 1, p. 7). This availability prevented any immediate liquidity crisis, and presumably reflected Cuba's strong record of repayments, continued belief in the Soviet guarantee, and perhaps European desire to maintain Cuban imports during a period of contracting domestic demand.

The accumulation of debt was not unanticipated by Cuban officials, though even their most conservative estimates had not predicted the extent and rapidity of the fall in sugar prices, and thus the magnitude of the required debt. The National Bank explained in retrospect:

> Imports demanded by the country in the two areas [capitalist and socialist] could, of course, not be compensated with export earnings, nor was it the intention then, since the necessary infrastructure did not exist to generate among other things, increased foreign exchange earnings . . . the only alternative the nation had in the face of the situation inherited from past governments and the political events in the sixties, was to step up capital formation through a massive flow of credit from abroad, unavoidably leading to a foreign debt, whose growth rate could initially be checked toward the end of the 1981–1985 period (that is, after a ten year period of maturity of the development program, which can obviously be regarded as a bare minimum). (BNC, August 1982, p. 16)

Following traditional development theory, and its 1970s incarnation — "debt-led growth" — Cuba was spending more than it earned in an effort to accelerate capital accumulation and facilitate economic restructuring. From 1975 to 1978, the ratio of investment to GDP was a high 23% (Mesa-Lago, 1981; Pérez-López, 1979, p. 80). The relative increase in investment per capita between 1969 and 1981 (306%) over consumption (175%) similarly reveals the stress on capital formation (calculated from CEE).

3. ECONOMIC DEVELOPMENT IN CUBA: SUGAR AND CAPITAL GOODS

The period in which the debt was incurred was marked by a renewed priority for industrialization. In 1981, 34.9% of investments were made in industry, compared with 16.7% in 1966 (Brundenius, 1984, p. 78). In the early years of the Revolution, an impulse toward self-sufficiency and independence had stimulated attempts at rapid diversification and import-substitution industrialization (ISI). These efforts never took off, but left Cuba with significantly decreased sugar earnings to pay for imports, required for both consumption and investment needs. By 1963, new plans adumbrated an expansion of sugar production. The earlier, more radical strategy was abandoned in wake of the apparently formidable costs it would entail before a less sugar-dependent economy could be achieved. The Revolution already faced severe shortages that threatened its immediate commitment to meeting basic needs. In the short run, at least, exporting sugar seemed more viable than diversification.

Geological and climatic conditions — hurricanes, droughts, disease, etc. — made reallocation of land from sugar to other less resistant crops a difficult task in Cuba as in the rest of the Caribbean (Hagelberg, 1985, p. 115). Sugar, in fact, appears to show considerable "comparative advantage" in the region (*Ibid.*; Boorstein, 1968, p. 199). Cuba's return to a sugar-based development strategy was predicated also on the Soviet Union's willingness to absorb the former US sugar quota. In the late 1950s, Kruschev had already set increased per capita consumption of sugar as a national goal; and Cuban sugar was hastily fit into an apparent socialist international division of labor (Boorstein, 1968, p. 200). Comparative advantage seemed to favor Soviet importation of sugar over domestic beet production (*Ibid.*, p. 199). And also an alternating cycle of harvests allowed Soviet imports and domestic production to complement each other, and thereby reduce the amount of idle refining capacity during the year, i.e., sugar could be imported and refined during the USSR's "dead season."[10]

It would be wrong, however, to assume an inherent contradiction between Cuba's efforts to augment sugar production, and those directed at expanding manufactures. The view that sugar is a staple unlikely to generate linkages within the economy, or provide any sort of "big push" (see Hirschman, 1981), is not well supported in the Cuban case. Instead the Cuban economy has witnessed a dramatic and intimately-connected development of its sugar and capital goods sectors. On the eve of the Revolution, no mechanization existed in either the cutting or loading of cane; and the most recent refinery had been completed in 1927. But the Revolution's very success in assuring social welfare, eliminating unemployment, and providing alternative jobs and education, resulted in a tight labor constraint on increasing sugar production that conditioned the timing of technological advancement. Laborers now found more promising opportunities than work in the sugar fields. As the new regime faced an increasingly severe shortage of canecutters, it simultaneously, and almost necessarily, ushered in a program of rapid mechanization. The initial failures of mechanization — for example, the high rate of harvester breakdowns — were compensated for by the mobilization of voluntary labor, which was often inefficiently transplanted from other sectors. The Swedish economist Charles Edquist (1985) concludes that a social consensus for change was fostered in socialist Cuba that was determinant of Cuba's successful mechanization strategy, one probably unique in the sugar-producing developing world (see Politt, 1981). The Revolution both destroyed erstwhile fears among labor that mechanization would eliminate needed employment, and provided a longer planning horizon for capital investment than private enterprise could sustain.

The results of Cuba's mechanization efforts are impressive. In 1985, two-thirds of all cane cutting was done by machines, while loading has been fully mechanized since 1978 (Brundenius, 1984, p. 74). Whereas 370,000 sugar workers were employed at the peak of the 1957/1958 season to harvest 5.9 million tons of sugar, 100,000 laborers in 1980/1981 obtained 7.4 million tons (*Ibid.*). A radical reallocation of labor from agriculture to industry and services thus occurred, representing a profound structural change. Official figures evince a reduction in employment in agriculture, fishing and forestry from 41.5% of the total in 1953 to only 21.9% in 1979 (*Ibid.*, p. 133).[11] And according to the Swedish economist Claes Brundenius, the percentage of agriculture in GDP declined from 18.1% in 1961 to 12.9% in 1981, while sugar (both agriculture and industry) dropped from 12.6 to 7.9% in the same years (*Ibid.*, p. 133). Mechanization gradually eliminated the rural labor shortage, and was accompanied by an impressive development of the skill and education level of the labor force which has contributed to development. The number of students in technical, secondary schools rose from 28,000 in 1970/1971 to over 300,000 in 1983/1984, while university enrollments climbed from 35,000 to over 200,000 during the same period (Brundenius, 1986).

Though efforts to diversify exports have generated only marginal gains in the relative earnings of non-sugar items, there are thus also indications that sugar production is simultaneously playing a smaller role in the overall economy. While sugar exports remain the motor for surplus-generation and foreign exchange, the economy appears increasingly driven by a growing industrial sector. In the late 1960s, Cuba had already provided the basic design for the world's most sophisticated and efficient harvester of green cane — *la libertadora*. Unable to produce the combine given the existing technological conditions, however, Cuba sold the patent to West Germany in an interesting reversal by which technology was transferred from the Third to the First World. Subsequently, a new combine harvester (KTP-1) was jointly developed by Cuban and Soviet technicians, which is now locally produced, including nearly all of the intermediate parts (Edquist, 1985, p. 134). Indeed, Cuba is now the world's largest producer of sugar-cane combines. Unfortunately, the technology was downgraded from the *libertadora* to permit local production, and thus is not yet of export quality (Edquist, 1985, pp. 134–135). The inability to market these harvesters represents an important missed opportunity for developing Cuban industrial exports, but may suggest a promising future direction.

The development of the sugar industry has effectively fostered both backward linkages, namely harvesters, and forward ones — the latter exemplified by the large energy needs now being met by bagasse, i.e., sugar cane pulp. For example, oil consumption in the production of sugar has now been completely eliminated (*Granma Weekly Review*, 10 February 1986, p. 3); and ethanol production may be another prospect for future diversification within the sugar industry. Also, the decrease in consumer goods from 39.1% of total imports in 1958 to 13.1% in 1982 suggests greater self-sufficiency, and an improved capacity for economic growth through concentration on capital and intermediate goods imports (CEE, 1972, p. 192; 1982, p. 329).

The real nub, however, is the increasing inability of sugar to fetch sufficient hard currency on the "free market." More than four-fifths of world production of sugar is now sold either internationally on preferential terms, or domestically to a protected market (Hagelberg, 1985, p. 100). The free market thus takes on a residual quality for most countries which largely accounts for sugar's low "world price" (*Ibid.*, p. 107). Free market prices for sugar in recent years have been consistently below the minimum cost of production, usually estimated at around 11 cents a pound. Rather than an equilibrium price and production level, a complex nexus of preferential (quota) and free markets exists in which profits are hampered in the latter by its residual quality, i.e., by the sale only of sugar remaining after quota purchases and domestic needs have been met, and of sugar frequently produced at a low marginal cost. All of Cuba's exports of sugar to the West, however, must be sold on the world market. A sign of Cuba's desperation for foreign exchange is that until the last few years, Cuba apparently gave the free market preference over USSR sales, despite the low world market price, and the increasing implicit subsidies offered by the Soviets (see below).

In the 1970s, commercial and government loans and serendipitous sugar booms supplemented meager hard currency export earnings for a strategy of economic development that relied on increased Western imports, and so-called "debt-led growth." The precipitous decline in sugar prices, however, exceeded even pessimistic forecasts, and was the penultimate blow to a development plan based on sustaining modest deficits with the West via foreign loans. The ultimate blow was the curtailment of loans.

4. THE 1982 DEBT CRISIS: CAPITAL ACCOUNT OUTFLOWS

By the end of the 1970s, Cuba had abandoned the strategy of "debt-led growth," restored external balance with the West, and sought to manage the existing debt in a prudent manner. In 1980, socialist Cuba registered its fourth and largest merchandise trade surplus yet with market economies, 174.8 million pesos, the result of increased sugar prices and continued concern to keep spending commensurate with national earnings. The 1980 current account in hard currency, however, remained marginally negative as a result of a deficit in services, i.e., interest payments, insurance, shipping, and tourism. But increases in tourism, Cuba's maritime transport, and its international assistance — civilian contracts for overseas skilled labor in construction, medicine, and education — continued to reduce the "invisible" trade deficit.[12] More importantly still, however, was a substantial growth in re-exports. And in 1981, despite a diminished merchandise surplus with market economies (44.8), the current account showed a hard currency surplus of 50.5 million pesos, the difference apparently issuing from new hard currency re-exports (BNC, August 1982, p. 33). In fact, since 1980, the value of hard currency exports has significantly surpassed figures for exports to the West. While a small portion of Soviet sugar purchases is sometimes paid in hard currency, the gap emanates mostly from new policies permitting the re-export of Soviet oil, which is reportedly sold, interestingly enough, to socialist, not market-economy countries (*Ibid.*; BNC, February 1985, pp. 13, 14, 35; Unctad, 1982, Anexo III, p. 12). As a result of these oil re-exports, Cuba has shown a large overall surplus in hard currency trade of 737 million pesos between 1981 and 1984 (BNC, December 1983, p. 32; BNC, February 1985, p. 14). And this surplus undoubtedly has been essential to Cuba's build-down, and effective management of the debt.

In terms of capital flows, however, the picture was more discouraging. After 1979, almost no fresh money entered Cuba. A fully-subscribed bond issue for 30 million French francs in 1979, and a syndicated loan for 150 million Deutsch marks in 1981 were even suspended following unexpected withdrawals by several banks. This new financial resistance may find partial explanation both in the fears generated by the need to reschedule Poland's Western debt, and the worsening US–Cuban relations which resulted from Cuba's military involvment in Africa, and Ronald Reagan's election as President. US pressures were applied to discourage Western banks from granting credit renewals or new capital (*Financial Times*, 23 March 1982). How causally efficacious US pressure was, however, is difficult to determine, as the contraction in credit also seemed to fit the pattern of restrictions in private bank lending to Third World debtors in general; and the bottom line may have been the low prices for sugar on the world market.

With interest rates soaring in 1980 and financial flows evidently drying up, Havana used some of its hard currency funds generated during the fleeting 1980 boom in sugar prices to reduce its Western debt. BIS-reported loans to Cuba had peaked in 1979 at 2.084 billion dollars, and fell to 1.834 billion by the end of 1980. Efforts to restore external balance, and limit the debt, however, did not exempt Cuba from a precipitous contraction in private bank lending the following year. In addition to capital outflows of over 100 million pesos from maturing medium- and long-term loans in both 1981 and 1982, 470 million pesos in short-term loans and deposits were withdrawn, i.e., not rolled over, in the 10 months from October 1981 to August 1982 (BNC, August 1982a, p. 43). Havana's vulnerability due to its unusually large percentage of short-term loans was demonstrated as commercial banks forced a sharp contraction in Cuban debt, reducing all BIS-reported loans to Cuba from 1834 billion dollars in 1980 to 869 million dollars in 1982. (Two-thirds of this reduction was in short-term obligations.)

As BIS-reported bank assets were thus more than halved, the National Bank balked at any further reduction of the debt. In a lengthy August 1982 report on the economy, the bank protested what it perceived as unfairly severe treatment; and asserted that if the current rate of withdrawals were continued,

> [it] would lead to the total reimbursement of Cuba's foreign debt from loans and deposits in convertible currency in about 33 months, an absurd assumption which no Third World country or even industrialized countries could meet, and whose very essence violated essential principles in international credit relations . . . [to do so] Cuba would have to increase her [hard currency] sugar exports by more than three million metric tons annually or stop all current imports from market economy countries. (BNC, August 1982a, p. 51)

Reserves were reduced from 338 million pesos in 1981 to only 96 million by the end of 1982 (BNC, December 1982b, p. 26). This reduction was required despite the surplus in trade with market economies in 1982 of 132 million pesos, and a current account surplus (the large outflows of interest payments notwithstanding) of 297 million pesos (BNC, December 1983 p. 32). Hard currency earnings continued to be boosted by re-exports, presumably Soviet oil. Already in 1981, re-exports ushered in some $200 million in hard currency, compared with 135 in 1980 and almost none in 1979 (Unctad, 1982, Anexo III, p. 12).

Partially in lieu of other possible, multilateral assessments of Cuba's debt situation and request for a rescheduling, Unctad prepared a report in 1982 on the Cuban economy. The study essentially reinscribed the Cuban government's view that Cuba faced overly restrictive financial treatment by international banks.

> The percentage reduction in the current Cuban debt was particularly large, and amounted to 25 percent, or three times the median reduction in the group of 28 countries whose debt fell in the last years. This seems almost paradoxical in light of the fact that ten of the other 27 countries were already behind on their payments at the end of 1981, while Cuba experienced no such problem until the withdrawals of short-term credit in 1982. (Unctad, 1982, p. 17)[13]

The apparent discrimination against Cuba reflected, at very least, the fact that the banks were not dealing with Cuba on an individual basis that recognized the country's strong financial record. Again, however, the greater debt reduction in Cuba may also find partial explanation in Cuba's large percentage of short-term debt which rendered the island particularly vulnerable to moves by the bankers to cut back loans and credit. Further, Cuba was in a weak bargaining position — with a lot to lose in terms of recently-regained financial credibility, and little capital with which to threaten default. Cuba's relatively early request for a rescheduling was issued only after an especially large reduction in the debt, and also following a prior restoration of balance in hard currency trade.

Cuba's immediate prospects for regaining, or increasing, hard currency export earnings, however, seemed dim. Above all, the sugar market was now indefinitely glutted, in wake of the accelerated use of high-fructose corn syrup and the increasing self-sufficiency worldwide.[14] Further, whereas the European Economic Community (EEC) had recently been a net importer of sugar, in the last decade it had become the world's largest exporter of sugar to the "free market" (BNC, February 1985, p. 26). The EEC, moreover, refused to join the International Sugar Organization and thus stymied efforts to control production and prices. At the same time, Cuba's expansion of non-sugar exports was very gradual — both in terms of agricultural diversification and of new industrial exports. Assured a domestic market and relying on CMEA technology, the production of competitive goods for export has proceeded more slowly than import substitution. Finally, the Reagan administration was tightening the screws of the Cuban embargo. The US strictly proscribed all imports with any intermediate products from Cuba, and renewed travel restrictions to Cuba which President Carter had allowed to expire in 1977.

Despite these obstacles, however, there were positive indicators also, and Unctad made a positive assessment of Cuba's prospects, and a pitch for rescheduling the debt:

> Though the possibilities for increasing exports are only moderately favorable until 1985, it is quite probable that they will improve substantially during the second half of the decade . . . as a result of increases in nickel, citrus, tourism, construction, and manufacturing . . . [But] for satisfactory econonomic management in the transitional period 1983–1985, it is essential to reschedule the bulk of Cuba's hard currency debt obligations due to mature in this period. (Unctad, 1982, p. 2)[15]

While Cuba has witnessed significant improvements in infrastructure, capital formation, and labor productivity, Unctad's expectations are optimistic and will depend on favorable international relations (tariff policies, tourism and access to international markets and technology), and the development of competitive industries for export. Certainly tourism shows considerable "comparative advantage," and potential for further growth — even without the natural and powerful US market. Despite the tightened US restriction on American travel, the total number of tourists rose to 200,000 in 1984; and they injected almost 70 million pesos of hard currency into the economy (BNC, February 1985, p. 15). Compared with 8 million pesos in 1978, the increase in tourism is impressive, and suggests the possibility for continued expansion in the future as facilities are improved, and more hotel rooms are made available (*Ibid.*, p. 37). Any relaxation in US–Cuban travel restrictions and international tensions would, of course, open up a much larger market.[16] Efforts to sell Cuban seafood products have also panned out. Fishing is now the second largest source of foreign exchange in hard currency exports, and garnered 103 million pesos in 1984.[17] In the short run, re-exports of Soviet oil (described below) will balance Cuba's hard currency accounts. In the long run, production and export of capital goods — for example, combine harvesters — will probably have to factor significantly into any achievement of external balance and self-sustaining growth (see Brundenius, 1985).

Despite Cuba's questionable short-term prospects, and perhaps contrary to US pressures, 13 creditor countries agreed in Paris in 1983 to reschedule the debt. The private banks followed. The debt was to be repaid in 10 equal semi-annual installments, the first in December 1985, the last in July 1990. The three-year grace period and 10-year maturity requested by Cuba were reduced to two-and-a-half and seven years, respectively. The interest rate was set at 2.25% above Libor — more or less the same high rate being charged in other reschedulings — and a 1.25% rescheduling fee. Short-term lines of credit would remain at their February 1983 level until 30 September 1984, with a continuing spread of 1.75% above Libor.[18] 1984 and 1985 packages rescheduled debt obligations on better terms — the latter with an interest rate at 1.50% over Libor; a 0.375% fee; and a 10-year maturity, with six years' grace (EIU, 1985, No. 3, p. 12). In 1984, short-term credit was re-extended also on improved terms — Libor plus 1.25%, with a 0.25% rescheduling fee (EIU, 1985, No. 1, p. 12). Notably, at the time of the original rescheduling, guidelines were set, such as debt service to hard currency income ratio, which, to some extent, were analogous to IMF conditions (Betancourt and Dizard, 1983, p. 6; EIU, 1985 Annual, p. 29).

Since 1982, the threat of a Cuban liquidity crisis seems to have subsided. The debt burden has eased in wake of reschedulings on increasingly favorable terms; a moderate extension of official bilateral loans; and a significant drop in the percentage of private, short-term loans, from roughly 40% of total debt in 1981, to only 23% in 1984. The question remains how Cuba has effectively managed its debt without traditional austerity measures or recession.

5. DEBT WITHOUT SOCIAL COSTS: SOVIET AID AND CUBAN SOCIALISM

In 1980, over 100,000 Cubans emigrated from the port of Mariel. That this massive exodus followed the sharp cuts in Western imports and economic recession in 1979 and 1980 might seem more than coincidental. Whether or not the retrenchment in Western trade and the Mariel evacuation were causally associated, it is clear that economic greivances weighed heavily on the minds of the emigrants, a large portion of whom were at the economic margins of society. Material discontent, moreover, was undoubtedly piqued by the opening of Cuba's doors to US relatives in 1979, by the American goods they brought, and the lifestyle they sought to portray. The traumatic image of Mariel and the concomitant specter raised by the Solidarity movement in Poland appears to have reinforced efforts in Cuba to increase the use and availability of material incentives, a trend which had already begun in the 1970s. An ominous fiscal crisis of the state thus loomed especially large as the onset of the debt crisis coincided with a stark recognition of the political-economic stakes involved in ensuring sufficient consumer goods, and continued economic reform.

In 1980, the minimum wage was raised by 14%. But to prevent a recurrence of money surplus (i.e., over goods and services available), prices had to be raised in 1981. Both the price of parallel market and essential goods (*por la libreta* — the Cuban system of rationing) rose significantly, the latter the first such increase since 1962. Food prices climbed 10 to 12% (Benjamin et al., 1984, p. 83). Personal savings were also encouraged. The newly established People's Savings Bank began offering up to 2% interest on deposits. In addition, Cuba promulgated its first major joint venture code. On the one hand, the law permits a large degree of foreign equity, 49%; and more in special cases, such as tourism, taxes are limited to a maximum of 30% of net profits; and a skilled, well-educated, and cheap (since the state pays the high "social wage") labor force is available (Schmidt, 1983).[19] Even Cuba's strong labor laws — for instance the difficulty in dismissing workers — have been rendered flexible in the case of joint ventures. On the other hand, Cuba's isolation from US markets may remain a significant disincentive. Recent agreements with Japan, Mexico, Panama, and Argentina, however, imply many favorable prospects for foreign investment. The Argentine company, Comarco, for instance, recently contracted with Cuba to build eight new hotels in Cuba's most important beach resort, Varadero, and thereby double existing capacity there (EIU, 1985, No. 2, p. 16).

Indeed, entering the 1980s, Cuba faced the likely constraints of a fiscal crisis of the state as sugar prices fell, and the economy was partially decentralized — the latter mitigating the government's almost complete control of the economy's surplus. But though shortages have been a chronic problem during lean years, the sharp edge of poverty and curtailment of basic social services seems to be absent in Cuban austerity measures. Nelson Valdés concludes:

> Apparently Cuban austerity measures are radically different from those found in capitalist countries. The overall social development of the population is not put in jeopardy. Instead, expenditures are reduced primarily in the productive sector (in industry and agriculture) . . . [The] revolution does not wish to affect the standard of living of the population, if possible. This is an ideological as well as political commitment. The recent experience in Poland may be present in the minds of the planners. Second, it is widely believed that investments made throughout the 1970s in the productive area should pay off in the future, so that it is possible to shift resources to consumption. Then, of course, there is a third element: Soviet economic assistance. (Valdés, 1983, p. 16)

Support for Valdés' assertion of a shift in expenditures from production to social needs can be detected in the figures from the state budgets. Though expenditures for production decline from 5,729.4 to 3,822.2 million pesos from 1981 to 1982, those allotted to housing and community services, education and public health, and other sociocultural and scientific activities each increased slightly (BNC, August 1982a, p. 53; BNC, December 1983, p. 9). In 1982 and 1983 these social services comprised 40% of the national budget, a significant increase over 33% in 1978 (*Juventud Rebelde*, 22 December 1977). And in 1985, social services even reached 46% of the budget (*Granma Weekly Review*, 20 January 1985).[20]

As the sugar industry approached nearly full mechanization, and large infrastructural needs were met, these capital investments could be tapered off. Similarly, the rapid decline in the birth rate following the initial postrevolution baby boom to a level equivalent to industrialized nations meant that education and other social *investments* could be decreased. In fact "non-productive" investments have declined sharply to only 15.6% of total investments in 1981, compared with 29.3% in 1962 (Brundenius, 1984, p. 78),[21] thus facilitating capital formation, even without increases in total investment.

If Cuban austerity measures really are "radically different," how exactly are they achieved? In addition to the successful implementation of previous investments discussed above, state control over, and redistribution of the national surplus (or profits), and massive Soviet aid seem to form the necessary and sufficient conditions of Cuba's austerity without threatening social costs. Until 1976, Soviet loans of almost $5 billion more than covered the large deficits in Cuba's trade with the Soviet Union (Mesa-Lago, 1981, p. 106). But the tied nature of this aid (export credits) diminished and rendered almost indeterminable the net benefit of Soviet balance of payments loans, the calculation of which would depend on the relative prices, quality, and available selection of Soviet goods, and also the eventual terms of repayment. Expressing this dilemma, Fidel Castro asked rhetorically in a 1979 speech, "Wouldn't it be better to get more towels and fewer television sets? Oh, if only that could be! But it is not a choice that can be made. The (CMEA) countries export to us products of which they have a surplus" (quoted in Theriot, 1982, p. 4). More importantly still to any analysis of Soviet assistance is the fact that until 1976 the cumulative subsidy from the Soviet price for Cuban sugar was less than what the island would have obtained if it was still receiving the US import price (Radell, 1983, p. 371).

Until the first oil shock, implicit Soviet–Cuban subsidies do not appear to have been significant. Since 1976, however, the Soviet price for Cuban sugar has far exceeded the US import price (see Table 4).[22] Not only have the convulsive price fluctuations of sugar been favorably smoothed out and buffered as in any quota agreement, but the price paid seems to entail a massive subsidy with which Cuba can buy Soviet oil, a commodity of unequivocal value. With each of the two waves of OPEC increases, implicit subsidies to Cuba from the USSR have increased dramatically, as they have, in fact, toward CMEA nations in general (Marrese, 1986). This apparently implicit sugar-for-oil deal has prevented in Cuba's case any sharp deterioration from pre-OPEC terms of trade with the Soviets, and more or less "sterilized" the Cuban economy from the oil price increases that have devastated most of the non-oil exporting developing world.[23] Despite the almost tenfold increase in oil prices between 1973 and 1982, and a concomitant decrease in world sugar prices, the purchasing power of Cuban sugar for Soviet oil was merely halved (calculated from unit values in CEE, 1975; 1979; 1982). Whereas one metric ton of sugar bought 74 barrels of petroleum products in 1973, a metric ton bought 36 barrels in 1982. On the world market in 1982, a ton of sugar could have been exchanged for only around six barrels of oil.[24] Thus, in contrast to Susan Eckstein's (1980) depiction of the capitalist constraints on socialist development, Cuba has, in fact, been unusually immune to the "world capitalist system," and able to limit Western trade when necessary, and capitalize on preferential Soviet prices.

A precise calculation of implicit Soviet subsidies to Cuba would require the creation of an economic model for overall terms of trade, external balances, and derived exchange rates that is far beyond the scope of this paper (see Marrese and Vanous, 1983). Apparently discounting the more complex mechanisms operating in CMEA terms of trade, however, the CIA calculates what could be considered an upper bound for estimates of Soviet aid. With the very extreme assumption that no significant opportunity costs are incurred in the tied nature of Soviet aid — i.e., that the prices and selection of Soviet products are as competitive and wide as on the world market — the CIA estimates subsidies based on the differences, *tout court*, between world market values of Cuban sugar exports and oil imports, and the amounts actually paid.[25] The figure that the CIA imagines ranges around $2.5 billion per year between 1976 and 1982 (CIA, 1984, p. 40), or about one-sixth of 1980 GDP (Brundenius, 1984, p. 39).[26]

The most straightforward problem with the CIA estimate is the unwarranted assumption that the favorable deal that Cuba gets in its sugar-for-oil swap is unmitigated by the selection and terms of trade for other products. In fact, petroleum products comprise only around one-third of Cuban imports from the Soviet Union (CEE). An almost equivalent portion is made up of machinery and transport equipment, Soviet items not usually thought to be of the highest quality. Writing under the auspices of the US Department of Commerce, even Theriot and Matheson (1979, p. 561) conclude that Cuba may be obligated to buy capital goods from the USSR that "because of quality or service deficiencies, are not readily saleable on the world market . . . If that is the case . . . what is apparently only a subsidy to Cuba in fact also accrues benefit to the USSR. Who gains the most is impossible to determine." Similarly, an old Eastern European joke tells of a man who returns home, and boasts to his wife, "I just sold our old dog in the market for 1,000 dollars." "Wonderful," she exclaims, "but what are you doing with those two ugly black cats." "Well," he explains, "in order to sell our dog for 1,000 dollars, I had to buy the two black cats for 500 dollars each."[27]

There is another problem with the CIA estimates. In a sense, it is inappropriate to compare the Soviet sugar price with the so-called world free market price, since sugar sold on the free market constitutes less than one-fifth of world sugar production. For political, planning, and protective purposes, most sugar is exchanged instead at higher quota prices. As one expert on Caribbean sugar asserts: "A better base (for calculating implicit subsidies) would be the weighted average of prices received from *all* markets, or an estimated equilibrium price" (emphasis added, Hagelberg, 1985, p. 122, fn. 11). In lieu of this weighted average, or equilibrium estimate, I have compared Soviet prices for Cuban sugar and average US import prices, since the gap between them became significant in 1976. Since then, the value of the implicit sugar subsidy has averaged around one-and-a-half billion dollars a year (Table 4). If we added the CIA's estimated implicit subsidy for oil imports, the total aid would average over $2 billion since 1976, or roughly one-seventh of 1980 GDP (CIA, 1984, p. 40). Again, however, this aid is given in non-convertible currency, and is tied to the purchase of Soviet goods, whose world market value is not determined. Furthermore, these aid calculations use official exchange rates which may be misleading, since the Cuban peso is not a convertible currency subject directly to the forces of the market, but rather is set by the government. Given the black market price of five or six pesos to the dollar, compared to the official rate of (roughly) 0.9 pesos to the dollar, the official peso rate appears overvalued *vis-à-vis* the dollar.

Thus, calculations which depend, *faute de mieux*, on the official exchange rate may indeed overestimate the dollar value of aid.

Another indicator that the magnitude of the subsidy is far less than it may seem is that until recently, at least — when oil re-exports began — the Cubans appear to have given the free market for sugar priority over Soviet sales (EIU, 1980, No. 3, pp. 7–8; 1985, No. 2, p. 12). If this is the case, hard currency from world market prices for sugar was, to some extent, considered more valuable than the far greater soft currency preferential price. Any convincing estimate of Soviet subsidies will thus have to assay carefully the complex nexus of hard and soft currency transactions, and market and socialist forces in which Cuba is operating. On the one hand, the central and unambiguous role of the sugar and oil exchange in Soviet–Cuban trade suggests clearly the importance of implicit subsidies. If valued at world market prices in most recent years Cuba could not have purchased even its petroleum products with its sugar earnings,[28] yet at Soviet prices petroleum products consumed only one to two-thirds of sugar earnings (Table 5). On the other hand, before an accurate estimate can be made of overall Soviet aid, a convincing determination will have to be made of the acutal world market value of Cuba's non-petroleum imports from the Soviet Union — a determination which may show the real purchasing power of yearly Soviet ruble transfers to be significantly less than the equivalent of $2 billion to $3 billion in hard currency suggested by a simple analysis of implicit subsidies for Cuban sugar and Soviet oil, and by Cuban balance of payments deficits (which expand Cuba's debt to the USSR). Above all perhaps, the problem is that the CIA's figure for Soviet aid to Cuba is averred by many writers without any qualification of the problematic methodology involved (see Zimbalist, 1983).

It should also be noted for comparative pur-

Table 5. *Cuban–Soviet trade in oil and sugar*

Year	Imports of Soviet petroleum products (million pesos)	CIA oil subsidy estimate ($ million)	World market value of Cuban sugar exports ($ million)	Soviet value of Cuban sugar exports (million pesos)	Total imports from USSR (million pesos)
1978	635	164	679	2,359	2,328
1979	743	381	818	2,263	2,513
1980	879	1,480	1,720	2,026	2,904
1981	1,139	1,657	1,190	1,939	3,234
1982	1,468	1,006	826	2,763	3,756

Sources: CEE (1982); CIA (1984); Table 4.

poses that by the CIA's calculations, Cuba would compete even with Israel for the world's most subsidized nation. Yet, upon closer analysis, the aid to Cuba pales by comparison. Whereas Cuba receives an estimated $2 billion in tied aid per year from the Soviet Union through implicit subsidies, with all the profound analytical problems in calculating the real value of such aid as outlined above, Israel received two billion dollars in outright cash transfers from the United States in 1985 (USAID, 1986, p. 172) — that is aid which is untied, and is in hard currency. Furthermore, on a per capita basis, Israel received in cash transfers in 1985 2.5 times the estimated average yearly trade subsidies that the USSR provides to Cuba.

Since the debt crisis, however, Cuba has garnered a new and formidable source of aid from the Soviet Union by being able to re-export Soviet oil for hard currency from what is purportedly saved from yearly quota. In a revealing commentary on the present logic of Soviet commitment, and the desperation of a developing country for foreign exchange, the 1985 BNC *Economic Report* frankly discusses how in the years 1983–85, 200 million pesos of sugar were bought on the free market "in order to meet as far as possible the commitments contracted with (the USSR)," and then sold to the Soviets for 1,328 million pesos. The 1,328 million pesos could then be used to purchase Soviet oil, a significant part of which was ultimately re-exported for hard currency (BNC, Feb. 1985, p. 35). The importance of this perquisite of Soviet relations is very clear; and undoubtedly it has been crucial to Cuba's successful management of its Western debt.[29] In wake of the abysmal levels of world sugar prices, re-exports of Soviet oil now constitute Cuba's leading hard currency earner. In fact, in 1983 and 1984, oil sales ushered in almost twice as much hard currency as sugar, representing, respectively, around 40% and 20% of exports in convertible currency (EIU, 1985, p. 4; BNC, February 1985, pp. 13–14). As a result, despite a large trade deficit to the West of 579.6 million pesos in 1984 (*Ibid.*, p. 13), the hard currency deficit was a mere 66.5 million pesos.[30]

Ironically for a nation heavily dependent on oil imports, however, Cuba will probably suffer significantly as a result of the current drop in oil prices, which has brought prices to less than a third of what they were just a few years ago, and thus attenuated what has strangely become the country's leading hard currency earner — oil re-exports.[31] Hence, both of the profound dimensions of Soviet aid — sugar subsidies and oil re-exports — have been closely tied to the rapid changes in oil prices over the course of the last decade, even more than to the fluctuations and decline in sugar prices; that is, both Soviet sugar prices and the reported earnings from re-exported Soviet oil have grown concomitantly with the OPEC price hikes. And the extent of implicit Soviet support relative to world market prices may now diminish significantly in the wake of fallen oil prices. Even if the preferential Soviet sugar prices remain high, the degree of subsidy will become far more questionable, as the value of Soviet products exchanged for Cuban sugar will no longer be as clearly prodigious as it was when the price of oil was at a peak. Also, the implicit subsidy for imported oil will decrease. The discount that the Soviet Union offers Cuba not only diminishes with each year after an oil price hike, but may be temporarily reversed in wake of rapid price decreases, as is currently the case, since though discounted for Cuba, the oil price is essentially tied to a five year moving average of world market prices.

Whether the recent high magnitude of total Soviet aid from implicit subsidies will represent a trend or a somewhat exceptional or extreme period will depend, it seems, on world market conditions (low sugar prices, high oil prices), and the USSR's willingness to continue compensating for declining terms of trade, and what Eckstein has called "capitalist constraints on socialist development" — constraints which have been especially severe in the context of the debt crisis, and also the continued US embargo against Cuba. Finally, it should be noted that Soviet aid to Cuba has not made the Cuban economy as privileged as it might seem relative to other developing countries, since Cuba suffers simultaneously from the unique problems of a complete economic blockade by the US. The Cuban government estimates that between 1959 and 1981, the embargo cost Cuba over $9 billion (BNC, August 1982, p. 12). And the long-term structural effects of exclusion from the US market undoubtedly make the opportunity costs much higher, as they prevent Cuba from maneuvering as efficiently as possible within the world capitalist and socialist markets, which is what, ideally I believe, Cuba needs to do.

It is also important to recognize that though Soviet aid is essential and formidable, Soviet subvention of the economy is not simply a dole that underwrites economic stagnation or invidious social reproduction. It would be wrong to discount what I believe is the other necessary condition, i.e., Cuban socialist polity, of Cuba's successful program of economic development, and effective management of the debt without traditional austerity measures. Accordingly, one writer responds to the CIA estimates of Soviet aid:

> Although $8 million per day is a gross overestimate of Soviet economic aid during 1980–1981, it should be pointed out that such aid levels are not unheard of outside the Soviet bloc. Chile for instance, a country with roughly the same population as Cuba, was running an average $10 million a day balance of payments deficit during 1981 which must have been supported by a corresponding aid program. Yet Chile can obviously not boast any of Cuba's economic and social accomplishments. (Zimbalist, 1983, p. 146)

It is also interesting to compare Cuba with Puerto Rico — the latter receiving from the US government in 1985, $4.8 billion in grants to agencies and individuals (US Census Bureau, 1985, pp. 76–77).[32] Puerto Rico thus obtains twice the upper bound estimate of yearly Soviet aid to Cuba, while it flows to a population roughly a third the size of Cuba's — in other words, Puerto Rico receives about six times the per capita aid that Cuba does. Yet Puerto Rico is hardly an analogous paradigm of growth with equity, and still depends fundamentally on a convulsive dynamic of incessant circular migration, or commuting, mainly between the island and New York City. Since the 1950s, almost a third of the population has migrated to the United States, while a quarter of those seeking work in Puerto Rico remain unemployed, and half the people live on food stamps and public assistance (Benjamin et al., 1984, p. 182). Though Cuban dependence on the USSR is clear, Soviet aid has financed development and greatly contributed to the success of Cuba's program of economic growth with equity (see Brundenius, 1984, pp. 109–124). The Soviets apparently provide the surplus needed for economic growth that was impossible for Cuba to obtain given its reliance on the chronically weak free market for sugar, and in wake of the OPEC oil increases.

At the same time, and in addition to Soviet aid, the social commitments and economic control in Cuba are essential to its program of socioeconomic development. Because the appropriation of surplus (i.e., profits) is not performed by the private sector, it can be distributed through the "social wage" and/or effectively invested rather than potentially dissipated on luxury imports, speculation, or unproductive investments, as in a country with a high return to capital, and great social inequality. Given the state's control over the means of production — and employment of about 94% of the labor force (Brundenius, 1984, p. 138) — the government has a virtual monopoly on the nation's surplus, and thus over development projects and equity concerns. Unlike the rest of Latin America where the state has difficulty obtaining needed revenue from taxes, and often relies on a limited number of more affluent private enterprises to distribute benefits to its employees, in Cuba the appropriate is direct, and hence guaranteed. Not only does the government thereby exert more control over the economy, but it can also effectively regulate and increase social equality. Greater equality may in turn increase popular support for the government, lend credibility to socialist ideology, and contribute to the potency of moral incentives — incentives which have been especially crucial to production and stability in the lean years in Cuba, and notably, a resource far more scarce in most developing capitalist societies. Conversely, during cycles of liberalization, the government may place less emphasis on egalitarian measures to permit greater use of material incentives.

Furthermore, the typical and formidable problems of capital flight are absent in Cuba. A clue to the magnitude of this "leakage" in other Latin American nations can be found in the figures for bank liabilities to (i.e., deposits of) foreign customers. Banks within the BIS-reporting range, for instance, report that at the end of 1982, Venezuelan-held deposits in bank accounts abroad ("liabilities") were equal to about half of commercial debt ("assets"), and Colombian deposits were equal to over two-thirds of outstanding loans. Guatemalan deposits abroad represented almost three times the size of commercial loans. Though some of these deposits are made by the state, the bulk of them reflect private capital movements. In Cuba, on the other hand, deposits in 1982 were equivalent to only 13% of commercial debt. Thus, if Soviet aid goes to Cuba, it cannot simply flow back to its source, as, for instance, much of US aid to Cental America does. Cuba faces neither guerrillas, devaluations, nor other forms of instability precluded by the strength of the state, and the contours of socialist polity.[33]

6. SOCIALIST DEPENDENCY?

Though marginal gains can be detected in comparisons of dependency in pre and postrevolution Cuba (see Leogrande, 1979), the country still remains in a seemingly classical mode of dependency — exporting a huge amount of sugar, mostly to the Soviet Union, and importing manufactured goods, oil, and a significant amount of food. Cuba's choice of a development strategy based on sugar has maintained structural dependence and a high degree of vulnerability. Both trade partner concentration with the Soviet Union and specialization on sugar play roles that seem to mirror prerevolutionary history. Yet evidence is scarce that current dependence on the Soviet Union shapes and delimits Cuban society in a manner analagous to Cuba's former dependence on the US. No one contends that Cuba has been subject to long-term capital outflows, deteriorating terms of trade, or extraction by the "core." And dependence on the Soviets entails neither foreign ownership of the means of production, nor, apparently, any intimate nexus with the economic vicissitudes of the socialist "center." Further, those socioeconomic indices normally associated with dependency — uneven development of town and country, and increased economic inequality — are conspicuously absent, indeed reversed, in Cuba today. Though the figures for health and education were relatively high in 1959 Cuba compared with the rest of Latin America, the signs of development and a high standard of living were rarely seen outside of Havana. In addition, in contrast to current figures, prerevolution health statistics were rather unreliable. Rates of infant mortality, for instance, must have been notoriously underreported given the limited provisions of social services and virtual absence of policlinics in rural areas (see Zimbalist, 1985, p. 8).

By all accounts, the urban–rural split has narrowed since the Revolution. And health and education standards now approach those of the developed world. Also, the unemployment that normally issues from late development in wake of a telescoped phase of light industry and recourse to capital-intensive technology, has instead been virtually eliminated. At the same time, the average real economic growth has paralleled that of the rest of Latin America between 1959 and 1980, and surpassed it since (Brundenius, 1984, p. 121). In fact, this statistic obscures the relative economic stagnation in Cuba during the 1960s, and the marked acceleration of growth following 1970. Since that time, the growth rate in Cuba has significantly outstripped that of the rest of Latin America (*Ibid.*), and the economy has remained robust even following the recession that has afflicted most of Latin America since 1980.

Though Cuba does not reflect the specific socioeconomic distortions described by dependency writers, however, the question of a peculiar, socialist dependency is not easily dismissed. Cuba depends on the Soviet Union as a vital, and irreplaceable market for its sugar, and, more recently, as the source of oil re-exports that ultimately balance the island's hard currency accounts. In the last instance, the dynamics of the Cuban economy are conditioned by political decisions made in Moscow over subsidies and trade patterns. And both the reasons for the Soviet's current generosity, and the particularly convoluted mechanisms by which the Soviets relay assistance, remain far from clear. Why not just give the Cubans the aid, whether in hard or soft currency? Perhaps, the symbolic value is what is at stake. The more circuitous methods obscure the extent of Soviet dependence and aid, and also render problematic, as we have seen, any rigorous estimate of them. Further, implicit subsidies and the high price of sugar can be portrayed as simply the paying of fair prices for primary commodities, and as anticipating the New International Economic Order. This portrayal serves Cuba's interest in appearing more independent than the island really is, and perhaps Soviet interests in veiling the aid from a potentially resentful population (Marrese, 1986).

The question remains whether the economic surplus obtained through the mechanisms of Soviet trade can generate self-sustained development or will delimit a structural dependence. There is certainly no *a priori* reason why a surplus, however obtained, cannot be used for sound investments and transformation of the economy. On the other hand, though the Soviets have not increased their domestic sugar production very significantly since the Cuban Revolution, and thus have become increasingly dependent on imports, it is unclear how much more sugar they can absorb.[34] One analyst believes even that the Soviets are already witness to a large black market of moonshine produced from Cuban sugar.[35] Yet the Cubans hope to increase production from eight to 12 million tons per year by 1990 (EIU, 1984, No. 3, p. 13). Given the surfeit of sugar in the world, and sugar's price on the free market below the cost of production, the heavy emphasis on sugar production which is implicit in Soviet subsidies may contribute to Cuba's dependence on the Soviet Union, and/or on an almost moribund free market for sugar.

In the longer run, non-sugar exports will have to increase if Cuba is to have self-sustaining growth and a more independent political economy. To the observer on the island witnessing the greater affluence on the store shelves today, and the substantial gains of most of the peasants and working class, the economy may appear as a showcase for socialism. But Cuba's balance of payments deficits and Soviet dependence betray a more precarious reality. While advancing toward full mechanization and high yields in sugar production, Cuba must also gradually decrease its monocultural dependence. Though the obstacles remain formidable, agricultural diversification and industrial development may proceed more rapidly given the extent of Soviet aid, and the strength of the Cuban state. Decreased reliance on sugar could contribute to a fuller realization of national goals of growth, equity, and independence.

NOTES

1. Unless stated otherwise, all Cuban debt figures are taken or calculated from Table 1.

2. The debt decrease would appear to be even greater given the devaluation of the peso. If the debt were converted to dollars using the official exchange rate, a reduction of 16% would be apparent.

3. As will be discussed below, however, the increases in oil prices also benefited the island through re-exports of Soviet oil — a boon which complicates any analysis of Cuba's sensitivity to the world market.

4. Unless stated otherwise, all Cuban balance of payments statistics with market and socialist countries are from Table 3.

5. By "implicit subsidies," I mean indirect aid resulting from the differences between Soviet prices and those in the free market.

6. In calling multilateral loans "more generous," I am referring to terms not amounts. Interest rates for 1970–78 are estimated by one author to average 50% less for World Bank loans — both IBRD and IDA — than for commercial Eurodollar loans, and to have average maturities four times as long (Stallings, 1981, pp. 208–209).

7. Theriot (1979) reports that group-of-ten bank claims accounted for 52% of outstanding CMEA debt in 1976 compared with only 16% for the developing countries.

8. My translation.

9. It should be noted that official parity between the dollar and peso was ended in 1972, reportedly in response to the high rate of inflation in the US. Official exchange rates set the peso equal to 1.09 dollars in 1972; 1.19 in 1973; 1.21 in 1974 and 1975; 1.22 in 1976; 1.26 in 1977; 1.33 in 1978; 1.38 in 1979; 1.41 in 1980; 1.28 in 1981; 1.20 in 1982; 1.16 in 1983; 1.13 in 1984; and was still falling in 1985 (CIA, 1984, p. 39; BNC, December 1984). Since the peso is not a convertible currency, its appreciation and devaluation reflect state policy and, only indirectly, market forces.

10. Personal conversation with Vladimir Treml (March 1986).

11. It is not clear whether the agricultural figures for contemporary Cuba cited in this paragraph include the private agricultural sector. If they do not, the figures may overrepresent the trend.

12. Unctad (1982), Anexo III, p. 15. Notably, service credits in 1978–81 far exceeded the combined revenue from tourism, freight, insurance, and interest payments, a discrepancy that seems to suggest over $100 million annually from internationalist assistance during these peak years. For a general discussion of Cuba's international assistance programs, see Susan Eckstein (1981).

13. My translation.

14. The trend toward increased beet and ersatz production is hardly a new phenomenon. The resultant fall in sugar prices in the 1920s left the post-1925 Cuban economy in a state of permanent stagnation (see O'Connor, 1970).

15. My translation.

16. A large increase in tourism, however, may have political costs for the regime as affluent visitors may cause resentment, and create new consumption needs.

17. If oil re-exports were considered, seafood products would be the third largest hard currency export. Without inclusion of oil re-exports, fishing represents 16% of hard currency export earnings — with them, it constitutes almost 10% (EIU, 1986, No. 1, p. 2).

18. An internal paper provided to me from a private banking source included this information, and was gleaned from various financial pages.

19. The state receives hard currency from the foreign enterprise, and pays the Cuban employees with pesos. By "social wage," I refer to the extensive provisions for social welfare (health, education, food) that are granted to all Cuban citizens.

20. These increases may also reflect both changes in relative prices over time, and the fact that enterprises have been required to bear more of their own expenses since the implementation of *calculo económico*. And thus the assertion of greater proportional spending on public consumption must be qualified.

21. Brundenius lists communal and personal sectors; science and technology; education; cultural activities; public health and social activities; and finance, insurance, and administration as the "nonproductive sphere" — in contrast to agriculture, forestry, industry, construction, transport, communication, and commerce, which are considered the "productive sphere."

22. This calculation for Soviet prices is based on the unit value of Soviet imports of Cuban sugar and, *faute de mieux*, the official exchange rate for converting pesos to dollars. The average US import price I use

comes from the US Department of Commerce's *Statistical Abstract of the United States* (various issues). According to the sugar desk at the Department of Commerce, the US imports by quota all sugar for consumption in the country; and the figures in the *Statistical Abstract* exclude any sugar bought on the free market that is refined and re-exported.

23. The largest price differentials from world market prices are those of sugar not oil. Although oil is discounted for Cuba, the price is basically tied to a moving average from the last five years of world market prices. Thus, the subsidy tends to be significant only during years immediately following large price increases, and was particularly high only from 1980 to 1982. (see CIA, 1984 p. 40).

24. I use a refiner acquisition cost for crude oil in 1982 of $33.55 dollars (American Petroleum Institute, *Basic Petroleum Data Book*, September 1985), and a conversion rate of 7.35 barrels per metric ton (Pérez-López, 1979, p. 282). The unit values of Cuba's imported petroleum products are calculated from CEE (1979; 1982). Sugar values are found in Table 3.

25. Aid estimates also include nickel subsidies, but these are insignificant (CIA, 1984, p. 40).

26. By reportedly adding development aid, trade deficits, and capital goods purchases from Moscow (i.e., the basis of Cuba's debt to the Soviet Union), the CIA further estimates total aid to Cuba in these years at $3.2 billion (CIA, 1984, p. 40). However, what the CIA means by the inclusion here of "purchases of capital goods from Moscow" is unclear to me, and the sorces for "public statements . . . concerning the amount of development aid" are not cited.

27. Michael Marrese relayed this joke to me.

28. This is apparent from the fact that the world market value in dollars of Cuba's sugar exports to the USSR for the years 1978 to 1982 (calculated from Table 3), save 1980, was little more than the amount in pesos of Cuba's imports of Soviet petroleum products (CEE, 1982); and that this is true despite the fact that substantial discounts from world market prices are made for Soviet oil in these four years (estimated by the CIA at an average of $800 million, CIA, p. 40), and that pesos are officially valued at an average of 1.33 dollars (see Table 4).

29. Interestingly, one author reveals that Bulgaria also was able to re-export Soviet oil, at least from 1979–82 (Marrese, 1986).

30. A large 7.4% growth in GSP in 1984 (BNC, 1985, p. 6) coincided — as the high growth rates in the mid-1970s did — with a substantial increase in Western imports. This high growth rate was justified by "the need to fulfill investment programs in order to rationally replace imports and promote new lines for export and to contribute to a greater overall economic efficiency in the country" (*Ibid.*, p. 36). The combined Western and socialist deficit in 1984 was the highest in Cuba's history.

31. The degree to which oil re-exports are ultimately a bookkeeping matter for informally transferring hard currency aid may then be revealed as new Soviet–Cuban policies may be necessitated, and new clues revealed about the nature of such policies.

32. I am indebted to Julie Feinsilver for this reference. In an unpublished dissertation manuscript, Feinsilver contrasts Cuba and Puerto Rico, and elucidates the greater provision of medical services in the former.

33. By the strength of the state I mean both the political-economic control of, and popular support for, the party, the government and the mass organizations.

34. Information on USSR sugar production can be found in the Food and Agriculture Organization Production Yearbooks of the UN (Rome). The area harvested has not increased at all, but yields per hectare have improved.

35. Personal conversation (March 1986) with Vladimir Treml, author of *Alcohol in the USSR* (Durham: Duke University Press, 1982).

REFERENCES

Banco Nacional de Cuba (BNC), *Economic Report* (Havana: August 1982a).
BNC, *Informe Económico Trimestral* (Havana: December 1982b).
BNC, *Quarterly Economic Report* (Havana: December 1983).
BNC, *Quarterly Economic Report* (Havana: December 1984).
BNC, *Economic Report* (Havana: February 1985).
Bank for International Settlements (BIS), *Maturity Distribution of International Bank Lending* (1978–85).
Benjamin, Medea, Joseph Collins, and Michael Scott, *No Free Lunch* (San Francisco: Food First, 1984).
Betancourt, Ernesto, and Wilson Dizard III, *Castro and the Bankers: The Mortgaging of a Revolution* (Washington: Cuban American National Foundation, 1983).
Boorstein, Edward, *The Economic Transformation of Cuba* (New York: Monthly Review Press, 1968).
Brundenius, Claes, *Revolutionary Cuba: The Challenge of Economic Growth with Equity* (Boulder: Westview Press, 1984).
Brundenius, Claes, "The role of capital goods production in the economic development in Cuba," Paper presented at the Research Policy Institute, University of Lund, Sweden (29–31 May 1985).
Central Intelligence Agency (CIA), *The Cuban*

Economy: A Statistical Review (Washington, DC: CIA, 1984).

Comité Estatal de Estadísticas (CEE), *Anuario Estadístico de Cuba* (Havana: 1972–82).

Eckstein, Susan, "Capitalist constraints on Cuban socialist development," *Comparative Politics* (April 1980).

Eckstein, Susan, "Structural and ideological bases of Cuba's overseas programs," *Politics and Society* (Fall 1981).

Eckstein, Susan, "Revolutions and the restructuring of national economies: The Latin American experience," *Comparative Politics* (1985).

Economist Intelligence Unit (EIU), *Quarterly Economic Review of Cuba, Dominican Republic, Haiti, Puerto Rico* (London: EIU, 1973–86).

Edquist, Charles, *Capitalism, Socialism, and Technology: A Comparative Study of Cuba and Jamaica* (London: Third World Books, 1985).

Hagelberg, G. B., "Sugar in the Caribbean: Turning sunshine into money," in Sidney Mintz and Sally Price (Eds.), *Caribbean Contours* (Baltimore: Johns Hopkins University Press, 1985).

Hirschman, Albert, "A generalized linkage approach to development, with special reference to staples," in *Essays in Trespassing: Economics to Politics and Beyond* (New York: Cambridge University Press, 1981).

Leogrande, William, "Cuban dependency: A comparison of pre-revolutionary and post-revolutionary international economic relations," *Cuban Studies*, Vol. 9, No. 2 (July 1979).

Marrese, Michael, "CMEA: Effective but cumbersome political economy," *International Organizations* (Spring 1986).

Marrese, Michael and Jan Vanous, *Soviet Subsidization of Trade with Eastern Europe — A Soviet Perspective* (Berkeley: Institute of International Studies, 1983).

Mesa-Lago, Carmelo, *The Economy of Socialist Cuba* (Albuquerque: University of New Mexico Press, 1981).

Mesa-Lago, Carmelo, and Jorge Pérez-López, "Imbroglios on the Cuban economy: A reply to Brundenius and Zimbalist," *Comparative Economic Systems*, Vol. 27, No. 1 (Spring 1985).

O'Connor, James, *The Origins of Socialism in Cuba* (Ithaca: Cornell University Press, 1970).

Pérez-López, Jorge, "Sugar and petroleum in Cuban–Soviet terms of trade," in Cole Blasier and Carmelo Mesa-Lago (Eds.), *Cuba in the World* (Pittsburgh: University of Pittsburgh Press, 1979).

Politt, Brian, "Revolution and the mode of production in the sugar-cane sector of the Cuban economy," *Occasional Papers* No. 35 (Glasgow: University of Glasgow, 1981).

Radell, Willard Jr., "Cuban–Soviet sugar trade, 1960–1976: How great was the subsidy?" *Journal of Developing Areas*, Vol. 17 (April 1983).

Roberts, C. Paul and Mukhtar Hamour (Eds.), *Cuba: 1986 Supplement to the Statistical Abstract of Latin America* (Los Angeles: Latin American Center, University of California, 1970).

Schmidt, Patrick, "Foreign investment in Cuba: A preliminary analysis of Cuba's new joint venture law," *Law and Policy in International Business*, Vol. 15, No. 2 (1983).

Stallings, Barbara, "Euromarkets: Third World countries and the international political economy," in Harry Makler, Alberto Martinelli, and Neil Smelser (Eds.), *The New International Economy* (Beverley Hills: Sage Publications, 1982).

Theriot, Lawrence, "Communist country hard currency debt in perspective" in *Issues in East–West Commercial Relations: A Compendium of Papers* submitted to the Joint Economic Committee, Congress of the United States (Washington, DC: US Government Printing Office, 1979).

Theriot, Lawrence, and JeNelle Matheson, "Soviet economic relations with non-European CMEA: Cuba, Vietnam, and Mongolia," in *Soviet Economy in a Time of Change* (Washington, DC: Joint Economic Committee, October 1979).

Unctad, *Cuba: Evolución económica reciente y perspectivas futuras* (UNCTAD/MFD/TA/21, November 1982).

USAID, *AID Congressional Report Fiscal Year 1987*, Annex II, Asia and the Near East (Washington, DC: USAID, 1986).

US Bureau of the Census for the OMB, Department of Commerce, *Consolidated Federal Funds Report Fiscal Year 1985*, Vol. 1, County Areas (Washington, DC: US Government Printing Office, March 1986).

Valdés, Nelson, "The Cuban economy and social developments in the 1980s," Paper presented to the International Conference of the Friedrich Ebert Stiftung on "Cuba in the 1980s" (Bonn: 1983).

Zimbalist, Andrew, "Soviet aid, US blockade and the Cuban economy," *Comparative Economic Studies*, Vol. 24, No. 4 (Winter 1983).

Zimbalist, Andrew, "Twisting statistics on Cuban health," *Cubatimes* (January–February 1985).

The Contributors

Claes Brundenius is director of the Research Policy Institute at the University of Lund, Sweden. He is former president of the Scandinavian Latin American Studies Association and has authored three books on Latin American development, most recently *Revolutionary Cuba: The Challenge of Growth with Equity*.

Alexis Codina Jiménez is professor and director of graduate studies in economics at the University of Havana. He is author of several books and articles on the Cuban economy and serves on the editorial boards of *Finanzas y Crédito* and *Economía y Desarrollo*.

Susan Eckstein is professor of sociology at Boston University. She is author of *The Poverty of Revolution: The State and Urban Poor in Mexico* and *The Impact of Revolution: A Comparative Analysis of Bolivia and Mexico*. She has written numerous articles on the outcomes of revolutions in Latin America and is currently completing a book on the Cuban revolution and editing a book on protest and resistance movements in Latin America.

Carl Henry Feuer is associate professor of political science, State University of New York, College at Cortland. His research focuses on the political economy of sugar production in the Caribbean, and he is the author of *Jamaica and the Sugar Worker Cooperatives: The Politics of Reform*.

Linda Fuller is assistant professor of sociology at the University of Southern California. She has published several articles on workers and unions in Cuba and is completing a book manuscript entitled *The Politics of Workers' Control in Cuba, 1959–1983: The Work Center and the National Arena*.

José Luis Rodríguez is deputy director of the *Centro de Investigación de la Economía Mundial* in Havana. He has written several books and dozens of articles on the Cuban economy and is a member of the editorial board of *Economía y Desarrollo*.

Sarah M. Santana works in epidemiology at the Center for Population and Family Health and the G.H. Sergievsky Center at Columbia University. She has done research in cancer mortality, reproductive epidemiology, and the health of Hispanic and Chinese minorities in New York City. She is presently investigating the determinants of the decrease in infant mortality in Cuba during the 1970s.

Jean Stubbs works with the research staff of the National Association of Small Farmers (ANAP) in Cuba and the Institute for Development Studies in Sussex, England. She is the author of *Tobacco on the Periphery: A Case Study in Cuban Labor History 1860–1958*.

Richard Turits received a masters degree in International Relations from Yale University and currently holds a Fulbright grant for collaborative research in the Dominican Republic. He plans to complete a doctorate in political science at Yale University.

Gordon White is a fellow at the Institute of Development Studies, University of Sussex, Brighton, England. He is a political scientist specializing in economic reforms in socialist countries, with particular emphasis on China.

Andrew Zimbalist is professor of economics at Smith College. He has written widely on the Cuban economy and has published several books in the fields of comparative economic systems and Latin American development, and he is editor of the Westview Press series *Political Economy and Economic Development in Latin America*.

Index

Absenteeism, 13, 142
Accumulation, 12, 28, 169; external sources, 17–18; strategies, 18–19
ACs. *See* Peasant associations
Advanced Peasants' Movement, 11
Agrarian reform, 10, 16, 25, 58; development and, 27–30
Agrarian Reform Laws, 34(n31); 1959, 27, 28–29, 51; 1963, 30, 51
Agricultural and Livestock Producer Cooperatives (CPAs). *See* Agricultural production cooperatives
Agricultural production cooperatives (CPAs), 58, 59(table), 65–66(n15), 71, 81(n3); tobacco, 53–55, 56(table)
Agricultural societies (SAs), 51, 54
Agriculture, 2, 14, 19, 30, 35(n50), 71, 79, 81(n1), 159, 161, 180(n14); commercial, 44, 49(table); cooperatives, 34–35(nn39, 40), 65; development of, 25–26, 31, 79; domestic production, 10–11; growth rates, 12–13, 41(table), 102(table), 103(table), 156; labor, 12, 51; machinery, 106–108; nationalization, 9, 10; production, 9, 15–16, 30, 35(nn 42, 54), 38(table), 39(table), 59(table); women in, 55, 57
Agro-industrial complexes, 70, 77, 82(n16)
Aid, 3–4, 19, 167, 175, 176, 177–178, 181(n26)
ANAP. *See* National Association of Small Farmers
Apparel industry, 89
Argentina, 167, 174
Assembly of People's Power. *See* *Poder Popular*

"Basic Organization of Labor," 135–136
Birth Rate, 16
Blockade. *See* Embargo
Brigades; production, 3, 15–16, 57, 58, 65(n9), 66(n20), 70–71; work, 14, 137–139
Bulgaria, 76

Cabaiguán, 62, 63; production cooperatives, 59(table), 60(table); tobacco growing, 43, 44, 51, 58
Campesinos. *See* Peasantry
Canada, 19
Capital, 18–19, 105, 172. *See also* Investment

Capital goods industry, 3, 112(n1); agricultural machinery, 106–108; development, 105–106, 170; production, 97–98, 99, 104–106
CCSs. *See* Credit and service cooperatives
Central Planning Board. *See* Junta Central de Planificación
China, 79
Claas Maschinenfabrik, 107
CMEA. *See* Council of Mutual Economic Assistance
Collectivism; economic, 158–160
Comisiones de reclamación, 142
Compulsory work law (1971), 13
Conflict resolution; in workplace, 141–142, 150, 151. *See also* *Consejos del Trabajo*
Consejos del Trabajo, 151; changes in, 147–150; conflict resolution, 145–147, 152(n16); function, 142–144, 150, 152(n17); organization, 144–n145
Construction, 102(table)
Consumption; promoting, 11–12, 16
Cooperatives, 65–66(n15), 81(n2); agricultural, 29, 31, 34–35(nn 39, 40), 59(table), 60(table), 65; organization, 15–16, 81(nn4, 5); sugar production, 71–72, 79–80; tobacco production, 51, 53–55, 58, 61, 63, 64(table); women in, 61–62. *See also* Agricultural production cooperatives; Credit and service cooperatives; Peasant associations
Corojo, El, 44, 51
Council of Mutual Economic Assistance (CMEA), 7, 19, 110, 113(n44), 156; sugar industry, 17, 176, 79; trade, 12, 166, 175
CPAs. *See* Agricultural production cooperatives
Credit and service cooperatives (CCSs), 51, 57, 58, 60(table)
CTC. *See* Cuban Labor Confederation
Cuban Communist Party (PCC), 123; First Congress, 135–136; planning role, 158, 159; Second Congress, 139
Cuban Labor Confederation (CTC), 142, 144
Cuban Land and Leaf, 44, 51
Cuban National Bank, 165
CUBATABACO, 58, 66(n22)

Debt, 4, 18, 156, 180(n2); increase, 17–18; origins of, 166–169;

reduction of, 172–173; western countries, 168–169. *See also* Debt rescheduling
Debt rescheduling, 5(n4), 165, 173–174. *See also* Debt
Decentralization, 5(n11), 21(n39), 81; decision making, 8, 54; health care, 116; planning, 15, 19, 66(n16), 159, 160–162; production, 70–71
Decision making, 7, 8, 54; central organization, 156–157; participation, 13–14, 158–160
Development, 1–2, 101, 170; agricultural sector, 25, 26, 27–28, 31–32, 58, 79; evaluation, 155–156. *See also* Economic growth
Directives for the Economic and Social Development of the 1976–80 Quinquennium, 135
Domestic Procurement Ratio (DPR), 108, 113(n48)
DPR. *See* Domestic Procurement Ratio

East Germany. *See* German Democratic Republic
Economic and Social Guidelines (1986–90), 139
Economic growth, 12, 85, 165, 170, 180(n12); domestic, 9–10, 17; exports, 8–9; external sources, 7–8; rates, 18, 166–167. *See also* Economy
Economy, 157, 171; evaluation of, 1, 155–156. *See also* Economic growth
Education, 21(n39), 142, 144, 170; health, 118, 120; labor force, 111(table), 112, 134; rural sector, 32–33
EEC. *See* European Economic Community
Eisenhower administration, 99
Ejercito Juvenil del Trabajo, 78
Embargo, 100; United States, 9, 99, 156, 173
Employment, 9, 16, 32, 170; rights, 143–144; sugar industry, 69, 78; tobacco industry, 57, 58, 66(n22). *See also* Income; Labor; Wages; Workplace
Engineering goods, 99, 100(table); growth rates, 102(table), 103(table); production, 104–105, 110
European Economic Community (EEC), 173
Export earnings, 7–8, 17. *See also*

185

Exports
Exports, 4, 33(n4), 109, 173; diversification, 18, 171; growth, 17, 171; production, 8–9, 11; sugar industry, 7, 9, 33(n2), 78–79, 171; to United States, 33(n1), 99. *See also* Export earnings; Re-exports; Trade

Family Code (1975), 62
Family physician program, 115, 116, 123; institution, 119–121; societal factors, 121–122, 124–125
Federation of Cuban Women (FMC), 66(n20); tobacco production, 57, 58, 62
Figueras, Miguel, 100–101
Fishing industry, 89–90
Five-Year Plans, 14, 15, 77; capital goods, 109–110; economic development, 100–104
FMC. *See* Federation of Cuban Women

GDP. *See* Gross Domestic Product
GDR. *See* German Democratic Republic
General Reform, 136–137
German Democratic Republic (GDR), 76, 109
Global Social Product. *See* Gross Social Product
Grievance Commissions. *See Comisiones de reclamación*, 142
Gross Domestic Product (GDP), 85, 100, 113(n48), 169
Gross Social Product (GSP), 10, 12, 18, 69, 85, 100, 101(table), 113(n17), 135, 139, 181(n30)
Gross value of output (GVO), 86, 87–88, 94(n12)
GSP. *See* Global Social Product
Guevara, Ernesto Ché, 99, 106, 129, 132
GVO. *See* Gross value of output

Hard currency, 4, 5(n4), 171–172, 173
Health care, 5(n10), 115, 125, 127, 179; accountability, 123–124; continuity of, 118–119; delivery system, 3, 116–118; rural sector, 27, 33; societal factors, 121–122. *See also* Family physician program; Primary care
Holguin machine plant, 107, 109
Housing, 33

Illiteracy, 33, 130
Imports, 9, 33(n4), 109(table), 168–169. *See also* Exports; Import substitution; Trade
Import substitution, 18
Incentives, 129, 135, 174; emulation, 132, 138; suppression of, 133–134; wage, 129–132
Income, 50(table), 80, 102; rural sector, 27, 32, 40(table). *See also* Employment; Wages
Industrialization, 28, 98, 99–100, 131. *See also* Industry
Industry, 2–3, 5(n9), 14, 136, 162(n3), 170; estimating growth, 86–89, 92(table), 93; growth, 12, 85–86, 89–90, 92(table), 102(table), 103, 155–156; net material product, 93–94(n2); product development, 90–91; production, 9, 11, 110. *See also* Industrialization
Infant mortality, 5(n10)
Infrastructure, 9, 31, 116
INPUD. *See* Instituto Nacional de Producción de Uténsiles Domésticos
INRA. *See* National Institute for Agrarian Reform
Institute of Internal Demand, 16
Instituto Nacional de Producción de Uténsiles Domésticos (INPUD), 161
Integral Brigades of Production, 137, 138
International Sugar Organization, 173
Investment, 31, 105, 170, 175; economic growth, 100, 101(table); 102(table), 103(table); foreign, 18, 19, 25, 174; income and, 11, 133; levels of, 18, 169; sugar industry, 76–77, 82(n11)
ISIC. *See* United Nations International Standard Industrial Classification of All Economic Activities

Japan, 19, 174
JUCEPLAN. *See* Junta Central de Planificación
Junta Central de Planificación (JUCEPLAN), 14, 21(n37), 156, 157–158, 162(n2)

Labor, 3, 12, 82(n19), 130, 150, 174; agriculture, 26, 27, 44, 54, 57–58, 66(n26); conflict resolution, 142–144, 145–147, 150–151; decision making, 13–14; discipline, 147–148, 149–150, 152(nn 18, 19); education, 20(n30), 111(table), 112, 134; engineering goods, 104–105; incentives, 129–132; organization of, 131, 134, 135–136; productivity, 134–135, 136; sugar industry, 77–78, 80, 82(nn 15, 17–19); tobacco industry, 43, 50(table), 51, 59(table); volunteer, 10, 65(n9), 78, 121, 132; wage reforms, 136–137; women in, 16, 55, 62; worker control, 142–143. *See also* Brigades; Employment; Unions; Wages
Land tenure, 25, 26, 29, 37(table), 46(table), 52(table); tobacco production, 43, 48(table), 51–52. *See also* Ownership
Latifundios, 26, 27, 29
Law for Agrarian Reform. *See* Agrarian Reform Laws
Law No. 8, 144
Law No. 32; function, 147–150, 153(n36)
Law of the Administration of Labor Justice (1962), 142
Law of the Organization and Financing of the *Consejos del Trabajo*, 144
Libertadora harvesters, 107, 108
Libor, 173
Licensing, 15
Liquidity, 172
Livestock production, 30, 40(table), 41(table), 45(table), 46(table), 49(table), 56(table)
Loans, 172, 180(n6); export earnings, 7–8; western countries, 17, 167

Management, 18, 82(n20); economic, 160, 161; worker discipline, 147–148, 150
Management Councils, 78
Manufacturing, 18–19. *See also* Capital goods industry; Engineering goods; Industry
Marketing, 181(n28); agricultural products, 2, 15–16, 35(n53); sugar industry, 79, 176–177
Maternal-child health care, 116
Medicine-in-the-Community, 116, 118, 119, 123
Mexico, 19, 174
MINAZ. *See* Ministry of the Sugar Industry
Mining, 89
Ministry of Agriculture, 30
Ministry of Labor, 134, 152(n5); conflict resolution, 142, 143
Ministry of Public Health (MINSAP), 116
Ministry of the Sugar Industry (MINAZ), 30, 70
MINSAP. *See* Ministry of Public Health
Mortality rates, 117–118(table), 119, 127
Municipal Tribunals. *See Tribunales Municipales*
Mutualist clinics, 116

National Assembly of People's Power. See *Poder Popular*
National Association of Small Farmers (ANAP), 30, 51, 66(n20),

INDEX

78, 81(n2); economic planning, 157, 159; tobacco production, 51, 53, 57, 58
National Center for the Scientific Organization of Labor, 134
National Institute for Agrarian Reform (INRA), 28, 30
Nationalization, 9–10
National Review Council, 143, 151
National School of Labor Judges, 142
National Tobacco Growers' Association, 44

Oil, 175, 176; re-exporting, 4, 172, 177, 180(n3), 181(nn 24, 31)
OPP. *See* Organs of Popular Power
Organization of American States, 167
Organs of Popular Power (OPP), 14. *See also Poder Popular*
Ownership, 11; agricultural land, 26, 31, 32, 34(nn 31, 38); reform, 28, 29, 30, 34(n31). *See also* Land tenure

Panama, 19, 174
Patricio Lumumba experimental tobacco farm, 51
PCC. *See* Cuban Communist Party
Peasant associations, 51, 54, 57. *See also* Peasantry
Peasantry, 27, 28, 35(n53); cooperatives, 30–31, 71; land ownership, 34(n38), 35(n45); population, 25–26; sugar production, 74–75. *See also* Peasant associations; Rural sector
People's Republic of China. *See* China
People's Savings Bank, 174
People's Stores. *See Tiendas del Pueblo*
Permanent Brigades of Production, 137, 138
Planning, 11, 99; agricultural production, 75–76; decentralization, 19, 66(n16), 159, 160–162; economic, 3, 8, 135, 155, 162(n4); participation, 13–14, 158–160; process of, 156–158
Planning and Management of the Economy, 135
Poder Popular, 14, 144, 146; economic planning, 157, 158, 159; health care, 116, 118, 123. *See also* Organs of Popular Power
Policlinics, 118–119, 120
Policy; agricultural production, 30, 31–32, 65(n14); associations, 123–124, 125(n19)
Popular Tribunals. *See Tribunales Populares de Base*
Poverty, 26–27
Pricing, 94(nn 15, 19), 134, 137, 176, 180–181; industrial growth and, 89–90, 91, 92(table), 93
Primary care, 116, 118, 120, 124. *See also* Policlinics
Private sector, 11, 15, 167; agriculture, 30, 31, 32, 66(n19); campesinos, 30–31; tobacco production, 59(table), 60(table)
Production, 54, 71, 73, 80, 103, 109(table), 152; agricultural, 2, 11, 31–32, 35(nn 42, 54), 45(table), 49(table), 52(table), 66(n17); agricultural machinery, 106–108; capital goods, 97–98; engineering goods, 104–106, 110; exports, 8–9; labor, 12, 134–135, 136; wages, 13, 16, 105, 131, 135. *See also* Agriculture; Industry
Production Assemblies, 78
Protectionism, 4
Provincial Tribunals, 148, 152(n14)
Publications, 89

Reagan administration, 19, 172, 173
Reciprocity Treaty, 98
Redistribution, 7; land, 29, 31, 33
Re-exports, 4, 5(n4), 180(n17), 181(n31), 172, 177
Reforms; collectivism, 158–159; economic, 15, 160–162; planning, 19; wage, 136–137. *See also* Agrarian reform; Agrarian Reform Laws; Redistribution
Regional Appeal Council, 143, 151
Rural sector, 27, 33, 44; education, 32–33; health, 27, 33; population, 25–26. *See also* Agriculture; Peasantry

Salaries. *See* Wages
San Luis; cooperatives, 59(table), 60(table); tobacco production, 43, 44, 51, 55, 58, 63
Santa Clara mechanical plant, 99–100, 106, 107
Santiago Rodríguez state tobacco farm, 51
SAs. *See* Agricultural societies
Scientific Organization of Labor, 131
SDPE. *See* Sistema de Dirección y Planificación de la Economía
VII Congress of Cuban Workers, 134
Sharecroppers' and Tenant Farmers' Association, 44
Sharecropping, 44
Sistema de Dirección y Planificación de la Economía (SDPE), 14–15, 16, 70, 136, 148, 156
60th Anniversary of the October Revolution plant, 107, 108
Social security, 32
Social Security Laws, 32
Soviet Union, 19; aid, 3–4, 167, 176, 177–178; dependency, 156, 178–180; price subsidies, 90, 94(n24); sugar industry, 17, 76, 78, 79, 180–181(nn 3, 5, 22); trade, 9, 175–176
State Committee of Labor and Social Security, 134
Sugar industry, 11, 19, 21(nn59, 64), 32, 49(table), 81(n10), 82(n23), 98–99, 101, 161, 170, 181(n23); cooperatives, 56(table), 71–72, 81(n5); development, 75–76; economic dependency, 171, 178–179; exports, 7, 9, 18, 33(n2), 78–79; investment, 76–77, 81(n11), 133; labor, 77–78, 80, 82(n15); land use, 26, 46(table); levels, 38(table), 45(table); mechanization, 73–74, 81(n7); problems, 74–76; production, 4, 41(table), 72–73, 80, 81(n6); state management, 69–70; subsidies, 19, 156; trade, 168, 173, 175, 180–181(n22); women in, 63, 82(n14)
System of Economic Management and Planning. *See* Sistema de Dirección y Planificación de la Economía

Technical schools, 112(table)
Technology, 18, 53, 97–98, 167, 171
Tertiary care, 121
Tiendas del Pueblo, 32
Tobacco industry, 33(n2), 45(table), 58, 63, 65(nn 10, 12, 13); cooperatives, 53–55, 56(table), 64(table), 65; land reform, 51, 53; land tenure, 43, 46(table), 47(table), 48(table), 52(table); value, 49(table), 50(table); women in, 57, 66(n22)
Tourism, 173, 174, 180(n16)
Trade, 3, 4, 17, 80, 101; CMEA, 12, 17, 76, 79, 80; deficits, 168–169; economy, 166–167; Soviet, 9, 175–176, 177, 179; surpluses, 171–172; western countries, 167–168. *See also* Exports; Imports; Re-exports
Training. *See* Education
Tribunales Municipales, 146, 148, 152(nn12, 14)
Tribunales Populares de Base, 152(n12)

Underdevelopment, 26–27
Underemployment. *See* Employment; Unemployment
Unemployment, 18, 33(n3)
Union of Communist Youth, 123
Unions, 13, 78; conflict resolution, 144–147; role, 145, 150–151; strength, 148–149, 151–152
United Nations International Standard Industrial Classification of All Economic Activities (ISIC), 104,

112(n1)
United States, 33(n1), 172, 174; embargo, 9, 99, 173; sugar industry, 25, 98
USSR. *See* Soviet Union

Wages, 105, 130, 138, 152(n21), 174; as incentives, 129–132; organization of, 134, 136; productivity, 13, 16, 134, 135; reform, 136–137; sugar industry, 78, 82(n17)
WEFA. *See* Wharton Econometric Forecasting Associates
Wharton Econometric Forecasting Associates (WEFA), 85; industrial growth estimates, 86–89, 94(n12)
Women, 122(table); agriculture, 55, 57, 59(table), 60(table), 61; cooperatives, 61–62, 65; health care, 116, 122; labor force, 16, 66(n20); liberation of, 80–81; sugar industry, 63, 77, 82(n14); tobacco production, 58, 61, 66(n22)
Work Councils. *See Consejos del Trabajo*
Workplace, 141; conditions of, 130, 133; management rights, 147–148; worker control, 147–148

DATE DUE

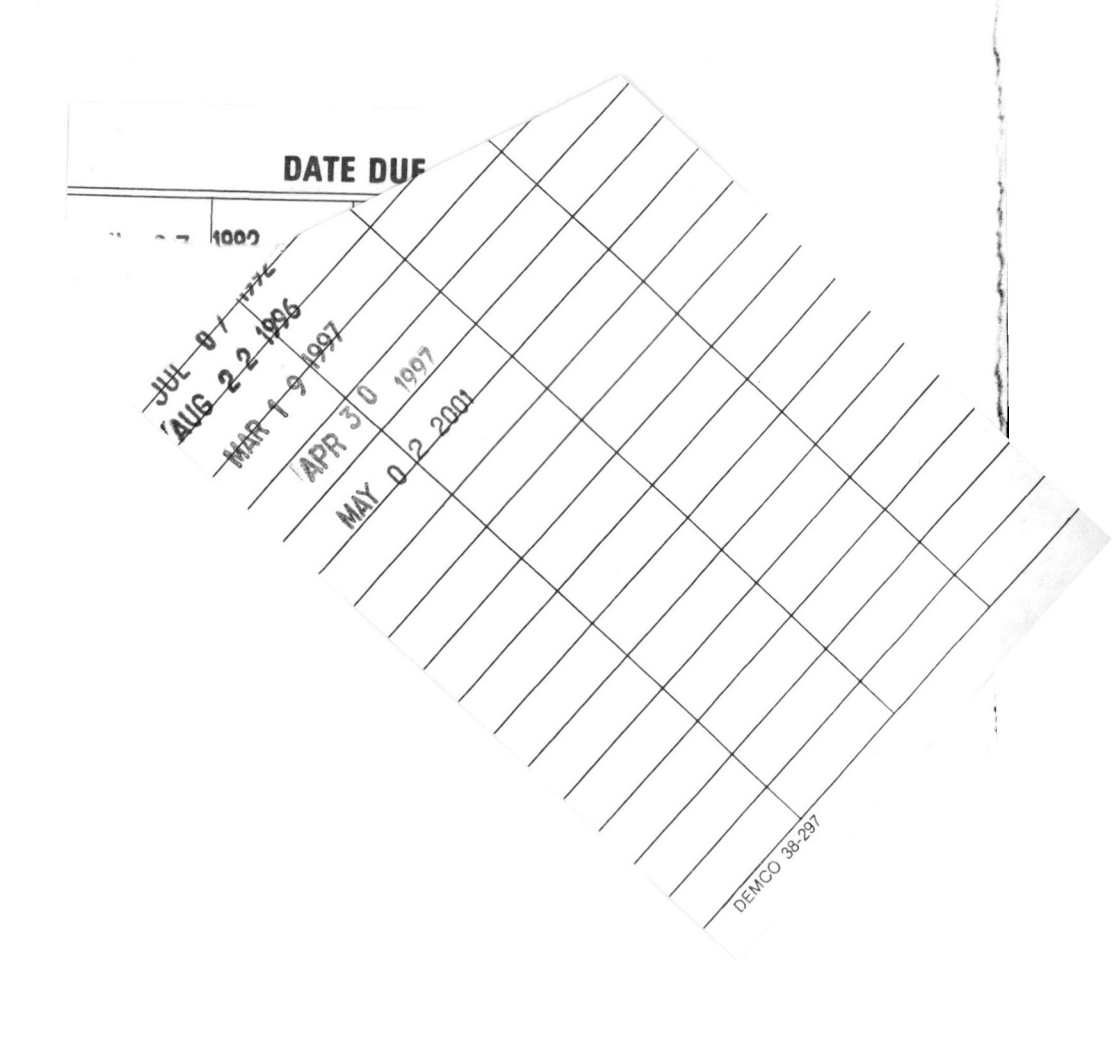

JUL 07 1992
AUG 22 1996
MAR 19 1997
APR 30 1997
MAY 02 2001

DEMCO 38-297